Praise

We're Not

"*We're Not Broken* is a lyrical mix of myth-busting, memoir, history, field interviews, and straight-up advice on how to better understand the autism spectrum, how to talk about it, and why it impacts every one of us."

—NPR

"Outstanding . . . Garcia's book uses rich storytelling and insightful reporting to uncover not only the long history of how autistic people have been mistreated but also how they continue to be ignored . . . *We're Not Broken* is exactly the book we need to lead the way in changing the autism conversation. It belongs on the shelf next to *NeuroTribes* as essential reading on autism and neurodiversity."

—*Washington Post*

"Garcia's strength is his ability to potently mix policy analysis (he covers the pros and cons of Medicaid Home- and Community-Based Services waivers, for example), reporting, and personal experience. This powerful account is packed with insight."

—*Publishers Weekly*

"*We're Not Broken* is a landmark book at a crucial moment in history, when autistic people are finally being recognized as the ultimate authority on their own lives. Surveying the whole autism landscape—from federal policy to intimate relationships—with heart, insight, and wit, Garcia's book will inspire generations of people on the spectrum to realize their fullest potential."

—Steve Silberman, *New York Times* bestselling author of *NeuroTribes: The Legacy of Autism and the Future of Neurodiversity*

"A bold and progressive exploration of autism in America today, *We're Not Broken* is an instant classic. Whether it's demystifying policy or profiling the next generation of autistic leaders, Garcia's story is guided by a simple principle: let autistic people articulate their own needs. In that way, to read this book is to listen to them."

—Ron Fournier, *New York Times* bestselling author and journalist

"Eric Michael Garcia has written a powerful and immensely clarifying book. *We're Not Broken* is more than a book about politics, autism, and disability. It is an essential meditation on the many ways of being human."

—Zachary D. Carter, *New York Times* bestselling author of *The Price of Peace: Money, Democracy, and the Life of John Maynard Keynes*

"In this groundbreaking and important book, reporter Eric Garcia uses in-depth interviews, media reports, and the most up-to-date scientific research to write a new narrative about autism. Turning his investigative skills inward, he plumbs his own life as an autistic person to tell the definitive story of how such a common condition came to be so misunderstood. *We're Not Broken* is essential, humane, and necessarily provocative reading."

—Jill Filipovic, author of *OK, Boomer, Let's Talk: How My Generation Got Left Behind*

We're Not Broken

We're Not Broken

Changing the Autism Conversation

Eric Garcia

HARVEST
An Imprint of WILLIAM MORROW

WE'RE NOT BROKEN. Copyright © 2021 by Eric Garcia. Afterword copyright © 2022 by Eric Garcia. All rights reserved. Printed in the United States of America. No part of this book may be used or reproduced in any manner whatsoever without written permission except in the case of brief quotations embodied in critical articles and reviews. For information, address HarperCollins Publishers, 195 Broadway, New York, NY 10007.

HarperCollins books may be purchased for educational, business, or sales promotional use. For information, please email the Special Markets Department at SPsales@harpercollins.com.

A hardcover edition of this book was published in 2021 by Houghton Mifflin Harcourt.

FIRST HARVEST PAPERBACK EDITION PUBLISHED 2022

Designed by Chloe Foster

The Library of Congress has cataloged the hardcover edition of this book as follows:
Names: Garcia, Eric, 1990– author.
Title: We're not broken : changing the autism conversation / Eric Garcia.
Description: Boston : Houghton Mifflin Harcourt, 2021. |
Includes bibliographical references and index.
Identifiers: LCCN 2020051612 (print) | LCCN 2020051613 (ebook) |
ISBN 9781328587848 (hardcover) | ISBN 9780358578338 | ISBN 9780358578741 |
ISBN 9781328587879 (ebook)
Subjects: LCSH: Garcia, Eric, 1990—Mental health. | Autism. |
Autistic People—United States—Biography. | Autistic people—Biography. |
Autistic people—Social aspects.
Classification: LCC RC553.A88 G364 2021 (print) | LCC RC553.A88 (ebook) |
DDC 616.85/8820092 [B]—dc23
LC record available at https://lccn.loc.gov/2020051612
LC ebook record available at https://lccn.loc.gov/2020051613

ISBN 978-0-358-69714-5 (paperback)

24 25 26 27 28 LBC 7 6 5 4 3

FOR CONNIE AND DELFINA

Your love sustains me

Contents

Introduction

This story starts in Des Moines back in 2017. Ostensibly, I was there to cover the Polk County Democratic Party's steak-fry, but I was really there because Iowa is where the seeds for both corn and presidential aspirations are planted. That morning, I was interviewing Representative Cheri Bustos of Illinois. There were rows and rows of corn in the fields outside the window of the office where we met, just like I had seen on television. I was decked out in my best grizzled-campaign-reporter outfit of jeans, flannel, and sneakers with my reporter's notepad in hand. Bustos was in jeans, high-heeled boots, and a red jacket as if to emphasize her rural Midwestern bona fides and show that despite being a Democrat, she could capably represent a district that had voted for Donald Trump in 2016.

Moments like these are why I became a reporter. I spent my teenage years wanting to play guitar in rock-and-roll bands, traveling the world and finding adventure. But when my parents told me I needed to find a legitimate job, journalism seemed like the next best thing. I could still travel the country, meet new people, and have that kind of anti-authority attitude that I got from my thrash-metal albums.

I always try to mention common ground with anyone I interview, and in this case, I brought up that I am a native of Bustos's home state and a Chicago Cubs fan. Later, I mentioned a hypothetical single mom raising a kid with autism, interjecting "like myself." The interview continued as I asked her whether she was considering a run for president. She gave a good nonanswer, coy enough not to be perceived as naked ambition so early in the game but not flatly denying she was thinking about the White House either.

As we finished the interview, Bustos, who used to be a reporter herself, turned the tables and asked: "So what is it like to have autism?"

After being a bit befuddled—it's a question I hadn't considered—I gave a vague answer, but Bustos's question stuck with me. How could I explain what it is like to have this massively complex condition that has only really been in the public lexicon for seventy years without getting it wrong? How could I share my personal experiences while also conveying that they are not indicative of the entire autism spectrum? At the time, I had been thinking a lot about autistic people's humanity and how often it is overlooked by the broader public. But interactions with both Democrats in Congress, like Bustos, and Republicans made me realize how many politicians just want to get it right. I realized there needed to be a book showing exactly what it was like to be autistic across America.

As a political journalist, I've spent my career watching Congress, first at *National Journal,* then as a reporter at *Roll Call,* then as an editor at the *Hill,* and then as an editor at the *Washington Post.* From this work, I know most politicians don't act with malicious intent. They are often just as good as the information they receive, and that information serves as the basis for their spending and budget priorities. In Washington and any state capital, money equals policy. Most politicians and members of Congress are too busy to become experts on every topic, so they rely on other people who are knowledgeable about a subject to inform their opinions. If their constituents or their advisers provide bad information, it will lead to them crafting policy that is harmful or ignorant of people's needs.

On a more micro level, when well-meaning people—from researchers and clinicians to journalists and parent advocates—have only tropes as their source of knowledge, they wind up inadvertently hurting their autistic friends, family members, neighbors, loved ones, and coworkers. The lack of autistic people's input means that even the experts who moved forward conversations in one regard can perpetuate stigmatizing ideas about autism in another. This goes back to the beginning of America's understanding of autism. Leo Kanner, a researcher at Johns Hopkins University, published the first landmark study of autism in 1943: "Autistic Disturbances of Affective Contact." While Kanner's study can be credited with bringing an awareness of autism to the broader public, that same study birthed numerous insidious myths about the condition that still exist today. Notably, Kanner wrote that while autistic children were born with an "innate inability" to form typical social contact with people, he noticed that "there are very few really warm-hearted fathers and mothers [of autistic children]." He later told *Time* magazine that the autistic children he studied were "kept neatly in a refrigerator which didn't defrost," thus creating the problematic stereotype of "refrigerator mothers."

Kanner's words took hold and found a champion in Austrian-American psychologist Bruno Bettelheim, who further smeared parents by saying they wished their autistic child did not exist. Even though Kanner would later "acquit" parents of responsibility for their children's condition and denounce Bettelheim's book, the theory had already permeated the zeitgeist.

Similarly, while psychologist Bernard Rimland helped relegate the idea of refrigerator mothers to the dustbin of pseudoscience, he also gave credence to the most popular contemporary misconception about autism: the idea that mercury in vaccines causes autism.

I was a teenager when I first realized how easy it is to spread misinformation about autism. I was watching Larry King interview former *Playboy* model and television host Jenny McCarthy, and she repeated the false claim that vaccines made children, like her son, autistic. The interview was broadcast on CNN, so I presumed there had to be some merit to the idea. I even asked my mom if I had been vaccinated around the same time that they found out I was autistic. She was puzzled; she'd never thought there was a link between the two.

Much later, I learned that McCarthy's claims were false, but she had been given a major platform, not just on CNN but also on Oprah Winfrey's show. And somehow, decades later, the myths and falsifications about vaccines and autism persist. Years later, I was living in Washington, DC, and working as a reporter for *National Journal*. It was 2015; the Republican presidential primary was heating up, and I had been assigned to cover a debate-watching party at a bar on Capitol Hill. At one point, I turned my eyes to CNN as moderator Jake Tapper asked Donald Trump about his previous tweets speculating a link between vaccines and autism. Trump, being Trump, doubled down. "Autism has become an epidemic," he said, and he talked about the spike in autism diagnoses in the past thirty years. He told a story about a friend whose child received a vaccine "and a week later got a tremendous fever, got very, very sick, now is autistic."

Like many things Trump says, this was, at best, wildly ignorant of the facts and, at worst, a brazen lie. My disappointment continued when one of his Republican primary opponents, retired neurosurgeon Ben Carson (who should know better), gave Trump a pass. This, of course, would become commonplace once Trump was elected: Trump would say something outrageous that Republicans knew was not true, but they felt they had to defend him, and they sacrificed their own credibility to do so.

Around this time, I was working on a major essay that would serve as the basis for this book in which I argued that society should stop trying to cure autistic people and instead help autistic people live fulfilling lives. Seeing Trump spread such toxic and dangerous information showed me that there was still a deep misunderstanding about autistic people. There is too much mythology and not enough data, an unfortunate reality that desperately needs to be counteracted.

It would be easy to dismiss Trump's comments as bombast and compare them to the many conspiracy theories he later espoused during his presidency. But he was buying into what many other people already believed. (I'll take an in-depth look at how the vaccine myth was spun in chapters 1 and 5.)

But that's not the only myth about autism that has played out in the media, where autistic adults are often portrayed as dangerous. Most notably, there is the myth of the autistic shooter—after reports that the gunman who killed twenty-six people, including twenty children, at Sandy Hook Elementary School was on the spectrum, calls to the Autism Speaks hotline jumped 130 percent. As Peter Bell, an executive vice president at the organization, told *Time* magazine, people wanted "to put the blame squarely on the fact that the shooter may have had autism." Bell added that the "rush to put a label on the situation has caused significant harm already."

Five months previously, MSNBC host Joe Scarborough stated that the man who shot up a movie theater in Aurora, Colorado, was "on the autism scale," although he had no evidence to substantiate this outrageous claim. (What was even more galling was that Scarborough has a son, Andrew, who is autistic. He later apologized.) When there was speculation in 2015 that a shooter at Umpqua Community College in Oregon was autistic, a petition had to be started to take down a Facebook page called "Families Against Autistic Shooters."

In reality, there is no evidence to indicate a connection between autism and violent behavior. This harmful stereotype is fueled solely by speculation and misunderstanding. Though we may never fully comprehend the twisted motivations behind mass shooters, autism alone is never the cause.

While these are all extreme examples, what they make clear is that many people have no idea how autism works, and the false narrative contributes to the idea that autism is a menace that must be eradicated lest helpless autistic children grow up to become stone-eyed snipers.

Many times, the misinformation that is spread about autistic people is rooted in the incorrect theory that autism is a problem caused by larger social ills (as anti-vaccine activists believe) or that it is a threat to the general public (as is the case with the myth of autistic mass shooters). And often, these falsities are spread by people who claim proximity to autism, like parents and researchers. That is why it is so essential for autistic people to be included in all parts of the conversation and why the conversation itself needs to change.

According to a 2019 report from the Interagency Autism Coordinating Committee, the federal government's advisory body on the condition, 75 percent of all research spending goes toward finding the root causes of autism and the best way to "treat" autistic people. The report also showed that the U.S. government spent only 6 percent of its budget on "research to improve services and supports for people with [autism spectrum disorder]." This contrast in spending reflects this country's backward values; rather than learning how to best assist autistic people as they navigate the world, researchers focus on mitigating symptoms and finding a cure.

But autism likely can't be cured. And despite the misconception that autism primarily affects children, eventually, those autistic children grow up. As they do, they need proper support systems and assistance to live their best lives within society. It would behoove the United States, a nation that leads the world in research funding, to focus on what autistic people really need: to get a job if they are able and not to live in poverty if they can't find one; to avoid discrimination; to receive an adequate education; to live within the community they choose; to have access to adequate health care; and, finally, to be free to pursue fulfilling personal lives.

An estimated one in fifty-four children in the United States has autism spectrum disorder, according to the Centers for Disease Control and Prevention. It is much more difficult to determine how many autistic adults there are because most testing methods are designed for children. Nevertheless, another CDC study estimated that 2.2 percent of the adult population is autistic. This book seeks to counteract the misinformation out there about autism to show what reforms are needed to improve the quality of life for autistic people in this country.

And when it comes to dispelling myths, I'm beginning with myself. My very existence contradicts what many people believe about autism. First off, I am entering my third decade of life; I'm not the child many people associate with this condition. Second, I am a person of color, which challenges the common idea that autism mostly affects upper-class white people. Furthermore, my line of work as a journalist exists outside the binary of what people think autistic people can do: I support myself with a job outside of the STEM fields, and I have done so since I graduated from college.

It is because I challenge how people view autism that I believe I am uniquely qualified to write this book. Frequently, people like me are portrayed as "overcoming" autism, something that is viewed as a great achievement, but autism is an integral part of my identity, and as such, it has played a role in my success. Since I began reporting and writing about autism, I have met autistic people who are much more intelligent and capable than I am but who are languishing because they did not have the same opportunities, support structure, or just pure luck that I did. The world around them is a bigger impediment to them than their autism ever was.

This book profiles autistic people across the country, from Nashville to Michigan, from Pittsburgh to the Bay Area of California to the nation's capital. Each person's story will hopefully debunk various myths about autism. There are many: that autism is something that affects only white males; that autistic people with more support needs are somehow less human or valuable than autistic people who require less support; that autistic people should appear in the media only when they inspire hope or pity; that autistic people are incapable of holding jobs outside of the technology field, of having relationships, or of living independently; that autistic people cannot advocate for themselves.

A tenet of journalism is that the writer is not the story, and I know that my story alone is not indicative of the story of autism in America writ large. I was born and was diagnosed at a time that the paradigm of autism was changing to include a broad spectrum. This was a few years before the pernicious myths about vaccines took the discourse about autism hostage. Disability rights activist Rebecca Cokley coined the term *ADA Generation* for disabled Americans who were in school when the Americans with Disabilities Act and the Individuals with Disabilities Education Act was passed, in 1990, the year I was born. In that same respect, autistic people who went to school after the ADA's passage are part of the Spectrum Generation, as they were often diagnosed when they wouldn't have been at any other time and were given their first tastes of freedom thanks to the ADA (I will discuss the Spectrum Generation in more detail in the next chapter). Throughout this book, I will include parts of my own narrative as a member of this Spectrum Generation and the ADA Generation where I feel it is necessary and appropriate to compare the experiences of the people I profile.

Last, I want to include some notes about language. First, it should be noted that as of 2013, autism diagnoses exist under the fifth edition of the American Psychiatric Association's *Diagnostic and Statistical Manual of Mental Disorders* (*DSM-5*) prescribed term of *autism spectrum disorders (ASD)*. But autism previously existed under various terms, including *infantile autism, Asperger's syndrome*, and *pervasive developmental disorder not otherwise specified (PDD-NOS)*. As such, throughout this book, there will frequently be mentions of those terms and others. The *DSM* and many labels are imperfect tools, and thus, when it comes to my sources, I will attribute their diagnosis based on how they self-identify. I sincerely hope that future generations will create terminology that more accurately describes autistic people's experiences.

I understand that because I am what many consider "high-functioning," I might not write accurately or empathetically about autistic people who require higher support needs. Thus, I include many people whom society might consider "low-functioning" to show that there are fewer differences between us than one might think. This book will also highlight the many difficulties that autistic people who are considered high-functioning still face. I want to break down the unhelpful and harmful binary that assumes people like me do not struggle while presuming incompetence in those who are nonverbal.

Even the terms *high-functioning* and *low-functioning* are considered inaccurate by some in the community, so I've made deliberate choices to use the terms autistic people prefer. Unless quoting other sources, studies, or published work, I won't use labels having to do with functioning. Instead, I'll refer to autistic people who, for example, cannot speak or who need 24/7 care as having "higher support needs." In the same respect, I identify those autistic people such as myself who do not require around-the-clock care, who might live independently, or who hold full-time jobs as having "lower support needs" unless they identify otherwise.

Many autistic people who require less support often live their lives undiagnosed; some who do get diagnosed choose not to disclose. I say this because I am one of those people. In my reporting, I learned how many LGBTQ autistic people exist and heard how many of them found parallels with accepting autism as part of their identity and revealing their sexual orientation or gender identity. In turn, I use the terms like *out* and *openly* to describe autistic people who incorporate their autism in their public identity. I will discuss autism and gender identity more in chapter 7.

This book will offer the most relevant data from peer-reviewed journals, government research, court cases, and media reports. In some cases, there will be citations of research that has been debunked, such as the theories about vaccines or Kanner's and Bettelheim's studies that blamed withdrawn parents for autism. I will also highlight areas where scientific researchers and autistic people profoundly disagree, such as over the concept of "extreme male brain" theory, which posits that autistic people's brains have more "male" traits, such as systemizing, and fewer "female" traits, such as empathizing (autistic people, particularly autistic women, strongly dispute this). I'll highlight when there is little to no information about a subject, such as the demographic makeup of the autistic adult population.

Some readers, especially parents of autistic children, may take issue with my authority here, since my experience may not be reflective of the broader autism community. I can't speak for everyone, but I've tried to include as many criticisms of my claims as possible to give them a good-faith airing before I explain why they are unhelpful, especially to their autistic loved ones.

Much of this book will speak about parents' misguided efforts to raise autistic children. This could lead to the misunderstanding that I am somehow anti-parent or critical of them. Nothing could be further from the truth—as you will see throughout this book, every opportunity I had in life came to me because of my family's dedication.

More than a critique, this book is an exploration of how misinformation and fear can drive even the most well-intentioned parents and people to make harmful decisions. I feature many parents who are dedicated to shifting people's attitudes about autism. Keivan Stassun, the parent of an autistic son and the director of the Frist Center for Autism and Innovation at Vanderbilt University, said something once that has stuck with me: "I would not change my son for the world, so I will change the world for my son."

Many neurotypical people who live with autistic people are committed to improving their loved ones' lives. I hope that parents and loved ones of autistic people take the lessons of this book to heart so they can be more effective allies.

Finally, I'd like to acknowledge that I'm lucky. At the time of writing this book, I rent a room in a house in Washington, DC, and I have an established career doing what I have wanted to do since I was fifteen. I've covered politics at one of the most prestigious newspapers in the United States during one of the most unprecedented times in our nation's history. Outside of work, I have great friends, a fulfilling life, and a family that supports me no matter what.

I've had the right opportunities at the right times. I have a mom who was able to stay at home full-time to raise me for my first twelve years and who did not buy into harmful myths about autism. I had professors who believed me when I said I was disabled and who were willing to accommodate me. I have had employers who were willing to work with me to make a positive work environment, which has allowed me to be a good employee. I have friends who love and accept me and gently guide me when I struggle with social cues.

Since I began writing and reporting about autism, I have met countless people who weren't afforded the same chances. Their educators doubted their need for accommodations. Their parents mourned their condition and subscribed to toxic tropes. They couldn't find work because employers were unaccepting. But none of that reflects who they are; it's a reflection of a world that penalizes them for not playing by its rules.

This book is a message from autistic people to help their parents, friends, teachers, doctors, and researchers see a side of autism that they may not have previously considered. It's also my love letter to autistic people, who will see that we've been forced to navigate a world where all the road signs are written in another language. Reporting about autism for the past half a decade has given me an identity and a home base, a place I did not previously have, so thank you to the autistic people who have welcomed my reporting on them.

We're Not Broken

1

"Don't Let Me Be Misunderstood"

POLICY

At Hyde Park Prime Steakhouse near Heinz Field, where the Pittsburgh Steelers play, the dim lighting, hors d'oeuvres, and staffers wearing campaign buttons corralling people to sign petitions all indicate one thing: There is a political event taking place. On this February afternoon in 2020, candidates—incumbent and challenger alike—are seeking the endorsement of the Steel City Stonewall Democrats, a group that represents lesbian, gay, bisexual, transgender, and queer Democrats in Pittsburgh.

LGBTQ rights are far from secure—if they were, the Stonewall Democrats would not need to exist—but this day, Jessica Benham is courting the group for its endorsement in hopes of taking another step toward acceptance. Benham, twenty-nine, is part of the Spectrum Generation: kids who were born and diagnosed around when the public's understanding of autism changed from viewing it as a form of childhood psychosis to seeing it as a broad spectrum that contains multiple variations. They were also the first group that benefited from legal protections and civil rights that had previously been denied to autistic people, however minuscule or piecemeal these statutes were.

They represent a shift in autism advocacy, one that places autistic people—rather than their parents, medical and psychiatric professionals, or researchers—at the center. They also advance the idea that autistic people and people with other developmental disabilities are not failures; they deserve acceptance from the broader public just as they are.

There have been at least two openly autistic state legislators in the United States before Benham. Briscoe Cain, a Republican from Texas, came out as autistic after he was elected to the state legislature in 2016. Democrat Yuh-Line Niou won her seat in the New York State Assembly and spoke about it to a student news website that same year. Benham would be the first autistic woman to be elected in Pennsylvania.

"I'm ready to make some history. And I don't know about you all, but where my heart is at is making some history for the Thirty-Sixth District," she tells the crowd. She doesn't immediately mention her identities. Rather, she rattles off goals relevant to Democrats in the audience, such as fighting for health care for all, combating the opioid epidemic that has ravaged Pennsylvania, and investing in infrastructure.

Only then does Benham mention she is autistic.

"I'm going to be the first out LGBTQ woman in the state legislature in Pennsylvania," she says to applause. "I'm also going to be the first openly autistic woman elected to a state legislature anywhere in this country." This leads to even more enthusiastic applause and occasional hollers.

Later, when I asked her about the sequence of her speech, she said that it was important to show that disabled people were not single-issue people.

"By the time I get to 'We're making history by being the first this or the first that'—which people seem to care about a lot more than I necessarily care about—people are riled up and then I don't have to come back down to be like, 'Let's talk about infrastructure spending,' which is my big thing," Benham tells me later while we are going door to door to gather signatures for her to get on the primary ballot.

A Contested Condition

Years and years of autism advocacy paved the way for Benham to get to this moment. Autism has likely existed as long as humans have, but it was a narrowly defined condition until very recently. Swiss psychiatrist Eugen Bleuler first used the term *autism* in 1911 to describe a symptom of childhood schizophrenia. This would be reflected when autism appeared in the first edition of the *Diagnostic and Statistical Manual of Mental Disorders,* in 1952, in the section on "schizophrenic reactions occurring before puberty." It cataloged the condition under an umbrella diagnosis of "psychotic reactions in children."

In 1968 in the *DSM-II,* autism again appeared as a symptom of prepuberty schizophrenia. *Infantile autism* finally appeared as separate from schizophrenia in the *DSM-III,* published in 1980. The condition had to have an "onset before 30 months of age"; the child had to display a "pervasive lack of responsiveness to other people" and exhibit "gross deficits in language development." If children could speak, they had to display speech patterns such as "immediate and delayed echolalia [parroting another person's speech]," peculiar language, and "pronominal reversal," which is referring to oneself in the second or third person but to others in the first. The diagnosis also required children to have "bizarre responses" to their environment, such as resistance to change, and "peculiar interest in or attachments to animate or inanimate objects." On top of that, to distinguish autism from schizophrenia, children needed to exhibit "absence of delusions, hallucinations, loosening of associations, and incoherence."

The *DSM-III* included diagnostic criteria for "childhood onset pervasive developmental disorder," another form of autism that in 1987 was renamed "pervasive developmental disorder not otherwise specified" and was used when children had impaired social interaction and communication skills "but the criteria are not met for Autistic Disorder, Schizophrenia, or Schizotypal or Schizoid Personality Disorder." The *DSM-III-R* revised the diagnostic criteria for "Autistic Disorder," requiring children to exhibit eight of sixteen different symptoms. Children (and it was mostly children) had to exhibit at least two traits in the "reciprocal social interaction" category—one trait that showed difficulty in communication and imaginative activity and one trait that showed "restricted" activities and interests.

The *DSM-IV* (1994) and the *DSM-IV-R* (2000) reduced the number of symptoms from eight to six. It also included a new diagnosis named for the researcher Hans Asperger, who had conducted some of the first studies on autism in Nazi-occupied Vienna. Unlike Kanner, Asperger theorized that autism was not a rare and narrowly defined disorder but a common condition that existed on a "continuum." His work was purportedly lost before British researcher Lorna Wing, a mother of an autistic daughter, argued that autism existed on a spectrum. (Since then, evidence has surfaced indicating the extent to which Asperger "actively cooperated" with the Nazis' child euthanasia program; he even referred children to a clinic where some of them died. This has removed much of the shine from his image.) Wing's husband, John, who spoke German, translated Asperger's work for her seminal 1981 article entitled "Asperger's Syndrome: A Clinical Account." Wing suggested that autism belonged in a "wider group of conditions which have, in common, impairment of development of social interaction, communication and imagination."

Like normal autistic disorder, an Asperger's diagnosis required the child to have social impairment, such as a lack of eye contact or difficulty developing peer relationships, as well as exhibit restrictive and repetitive behaviors. But unlike a child with autistic disorder, a child with Asperger's did not have a "clinically significant general delay in language," nor a "significant delay in cognitive development or in the development of age-appropriate self-help skills, adaptive behavior (other than in social interaction), and curiosity about the environment in childhood."

But nineteen years after the clinical introduction of Asperger's—which is to say, a year after kids diagnosed with Asperger's in 1994 could vote but two years before they could legally drink—the diagnosis for autism would again change. In 2013, the American Psychiatric Association placed autistic disorder, Asperger's syndrome, childhood disintegrative disorder, and PDD-NOS in the *DSM-5* under the umbrella term *autism spectrum disorder.* This new label for the condition is characterized by "persistent impairment in reciprocal social communication and social interaction" and "restricted, repetitive patterns of behavior."

All of these changes to the *DSM* led to a surge in autism diagnoses, which makes sense; if the definition of autism is broadened, it stands to reason that more people who went undetected in the past would get diagnosed.

And throughout the years, these changes in diagnostic criteria mirrored changes in public policy. In 1990, President George H. W. Bush signed the Americans with Disabilities Act. Congress barely mentioned autism during deliberations and it appeared nowhere in the text of the law. The law was far from the windfall that disabled people needed, but it was an instrumental in portraying people with disabilities as a group deserving of civil rights. Regulations from the ADA Amendments Act of 2008, which expanded the definition of *disability*, specifically named autism as a disability, and therefore, autistic people were protected by it.

The same year the ADA was passed, Congress reauthorized the Education for All Handicapped Children Act under a new title, the

Individuals with Disabilities Education Act (IDEA). The House of Representatives report on the legislation said that "autism has suffered from an historically inaccurate identification with mental illness" and that including autism in IDEA was "meant to establish autism definitively as a developmental disability and not as a form of mental illness." IDEA mandated that students with disabilities were provided with a "Free Appropriate Public Education," and now that specifically applied to students with autism. It also meant that schools had to report the number of autistic students they served, along with other disabled students, to the U.S. Department of Education. These improvements in knowledge and changes in policies had the potential to move American politics toward a more accepting model of autism.

I am a beneficiary of these changes and a member of this new Spectrum Generation. These sea changes happened as I was growing up; I was born in 1990, the year the ADA was passed and the changes to IDEA were made. Around the time the *DSM* would come to include Asperger's syndrome, my family was living in Wisconsin. It was there that clinicians asked my mom to bring me in for extra tests, an ordeal that ultimately led to my Asperger's diagnosis. I was a college student at the University of North Carolina when the *DSM-5* changed its criteria to put all variations of autism under the umbrella diagnosis of autism spectrum disorder (ASD). Much of my schooling from fifth grade onward was at private parochial Christian schools, but in my early years in public school, thanks to IDEA, I had special education services and tutors to help me with math, English, and handwriting. When I got to college, the ADA meant that I could have reasonable accommodations—like extra time on tests and separate locations to take them—which meant I never had to worry about my environment; I only had to worry about my studies. These policies were not charity; they allowed me, and scores of other disabled people, access to the same freedom that nondisabled Americans take for granted. It was my chance to cash the check of freedom guaranteed in the Declaration of Independence.

Before I ever learned I had a disability, politics fascinated me and became my special and myopic interest. My mom taught my sister, Stephanie, and me to read when I was three, and she constantly pressed us to learn about American history. My mom was adamant about us learning how to read at an early age and I particularly loved presidential history. As Mexican Americans, my parents firmly identified with the latter part of that term while feeling pride in the former.

My hunger for history led to my understanding that politics is essentially history in action. Today's newspaper articles become footnotes in tomorrow's textbooks. My dad has been an ardent Texas Republican since the Reagan administration and watches Fox News regularly, and my stepdad, Bob, carries his Democratic roots from his youth in Allentown, Pennsylvania, and loves Joe Biden. In this way, it makes complete sense that I wound up working in political journalism, covering campaign speeches in Iowa and Illinois, interviewing senators near the underground train that connects the office buildings to the main Capitol Building, and prognosticating how states would vote in future elections.

First my interest in politics was a hobby, then it was a career, but as I read more about the history of autism and disability rights, politics became intensely personal for me. I realized that I get to report in the halls of Capitol Hill because of the work of disabled activists who literally crawled up the steps of that very building to help pass the ADA. When people see me as an inspiration because I "overcame" my disability to graduate college and hold a job, I want to respond that the only things I overcame were the specific obstacles in front of me. I am a return on others' investment in policy. In the same way, every autistic person who languishes in classes or winds up in a group home or institution is not a reflection on a poor upbringing but rather a failure in policy.

I was a teenager when society, including politicians and celebrities, began to take notice of autism. It shaped how I saw myself in ways I did not fully grasp in the moment. While I had a vague understanding of myself as someone with Asperger's syndrome, I did not know what that meant, other than that it was some type of autism. As an aspiring guitarist, I cared more about learning how to write my own riffs that were as heavy as AC/DC's and Black Sabbath's. One day, when watching a show on VH1 Classic, a channel that played my favorite metal music on weekends, I saw a public service announcement with many of my favorite musicians rattling off harrowing statistics about a mysterious condition.

Tommy Lee, the drummer for Mötley Crüe, rocked a pink faux-hawk and opened the PSA warning "as many as one in a hundred and sixty-six children are diagnosed." He was followed by Roger Daltrey—the frontman of the Who, the first band I ever saw in concert—saying, "That's more than sixty-eight a day." The ad continued with Gene Simmons and Paul Stanley—replete with their kabuki makeup and Kiss regalia—saying that more people had it than "type one diabetes, childhood cancer, and cystic fibrosis combined." Robert Plant, who looked less like the sex-symbol frontman from Led Zeppelin and more like a *Lord of the Rings* character, said, "No one knows where it comes from."

The ad made teenage me wonder, *What kind of terrible affliction is this?* Finally, it culminated with Ronnie James Dio of Black Sabbath, the singer of the very band I aspired to emulate, saying in his deep baritone voice, "Help us put an end to autism."

Initially, I cradled that PSA closely. I felt these musicians who dominated my airwaves and my iPod had my back. I don't think I cared about being cured of autism then; what mattered more was that someone saw me. But now, I watch that ad and think about how corrosive it was to see Joe Perry and Steven Tyler of Aerosmith, whom I loved when I saw them live, say they wanted to wipe me out. I think about the irony of the fact that Kiss, a band that taught me that being aggressively bombastic can be a good thing, saw autism as a disease on par with cancer.

Rob Halford of the band Judas Priest was an openly gay man who adopted the regalia of studs and leather from gay bars in London that became the heavy metal "look" that so many macho metal dudes would copy. For me, he was the gateway to making a straitlaced church kid accept LGBTQ people. But while I imagine he would be deeply offended if anyone tried to "cure" him of homosexuality, there he was, advocating for a cure for what made me, well, me.

I don't deny that those musicians wanted to do some good with that ad. The prevailing wisdom at the time, influenced by many parents' advocacy groups, was that autism was a terrifying condition that required a full-frontal assault. Those musicians probably thought they were using their wealth and influence to help young kids. But their message was that autism is a death sentence, that it would be better if autism was eradicated. It is that spirit that makes autistic people seem like they are either objects of pity or "inspirational figures" who don't let their condition "hold them back," when it is often the world and its unwillingness to adapt that is much more of an impediment to them.

One of the other parts of that ad that I remember is Twisted Sister's Dee Snider looking into the camera and saying that autism was "the fastest-growing developmental disability in the United States." Of course, we now know that those numbers grew not because more kids were autistic but because the tools used for diagnosis were improving.

Around the same time, Republican Representative Chris Smith of New Jersey and Democratic Representative Mike Doyle of Pennsylvania cofounded the Congressional Autism Caucus. I caught Doyle campaigning for an endorsement at the same Stonewall Democrats endorsement event that Jessica Benham was at. When he made his case for their support, he mentioned cofounding the caucus. Later, Doyle told me that at the time, he noticed how passionate parents' groups were because their children were being institutionalized or were not getting the services they needed.

"It didn't matter whether we found out what was causing it," he said. "They were there, and they needed help."

NBC/Universal chairman and CEO Bob Wright and his wife, Suzanne, founded the organization Autism Speaks in 2005 after their grandson Christian was diagnosed with autism. The group's mission statement said the nonprofit was "dedicated to funding global biomedical research into the causes, prevention, treatments and a possible cure for autism."

The nascent nonprofit quickly gained national prominence when Congress passed the Combating Autism Act with bipartisan support. President George W. Bush, whose father had signed the Americans with Disabilities Act back in 1990, said, "I am proud to sign this bill into law and confident that it will serve as an important foundation for our Nation's efforts to find a cure for autism."

From the 1970s until the mid-2000s, much of the political advocacy for autism was done without the input of autistic people. The fact that autism existed on a spectrum and that autistic people could speak (either through their physical voices or via communication devices) and advocate for themselves was not widely known until the 1990s.

To understand the tension that exists today between parent advocacy and self-advocacy, one must first understand its origins. For much of the early twentieth century, medical and psychiatric experts dominated the conversation about autism. Leo Kanner conducted one of the first widely read studies about autism in 1943, but his study noted many of his subjects lacked "warm-hearted" parents. He later elaborated on this in a 1949 paper, saying that "many of the fathers remind one of the popular conception of the absent-minded professor who is so engrossed in lofty abstractions that little room is left for the trifling details of everyday living." He also wrote that "maternal lack of genuine warmth is often conspicuous in the first visit to the clinic" and that he'd observed only one mother fully embrace her son at her first visit to his clinic.

"They were the objects of observation and experiment conducted with an eye on fractional performance rather than with genuine warmth and enjoyment," Kanner wrote of the autistic children he studied. "They were kept neatly in refrigerators which did not defrost."

Kanner's words gave rise to the concept of "refrigerator parents," and Bruno Bettelheim, who led the Orthogenic School in Chicago, further popularized the concept. Bettelheim's book *The Empty Fortress* explicitly blamed parents for giving their children a condition, going as far as to say, "I state my belief that the precipitating factor in infantile autism is the parent's wish that the child did not exist." He called autism "a state of mind that develops in reaction to oneself in an extreme situation, entirely without hope."

Even worse, faulty research like this "helped keep autism *off* the agenda," as Claremont McKenna College political science professor John M. Pitney Jr. wrote in his book *The Politics of Autism.* Autism was framed as a personal concern rather than a collective one and thus didn't require broader policy solutions. Later, Kanner would absolve parents of blame, and Bettelheim's ideas would fall out of fashion. But the damage they did to parents who blamed themselves and their autistic children was immeasurable personally and politically, since it prevented autism from being part of the growing disability rights movement.

Thus, the early era of autism advocacy in the 1970s was meant as an act of taking back power for parents who had been blamed for their children's autism for years. And parents made some valuable net gains. Ruth Christ Sullivan, one of the founders of the National Society for Autistic Children (NSAC, later the Autism Society of America), lobbied for the condition to be included in the Developmental Disabilities Act, which was signed by President Gerald Ford. The second part of the legislation was titled "Establishment and Protection of the Rights of Persons with Developmental Disabilities" and said explicitly that "persons with developmental disabilities have a right to appropriate treatment, services, and habilitation for such disabilities" and placed autism under the umbrella of developmental disabilities.

"Those brief 3½ years since its passage have seen the establishment of many services for our children, often for the first time, usually where none existed before," Sullivan wrote in a 1979 article published in the *Journal of Autism and Developmental Disorders.*

But some of the parents who sought to recalibrate the scales would themselves tip them one way or another. Researcher Bernard Rimland, who cofounded NSAC and was a parent of an autistic son himself, helped challenge the toxic-parenting theory. But his organization Defeat Autism Now! also promoted treatments like casein-free and gluten-free diets and the use of vitamin B_6, as well as more dangerous crank theories, such as intravenous use of the hormone secretin.

As a result of parent advocates dominating the conversation, autistic people themselves wound up missing out on much of the disability rights movement that animated the twentieth century.

"One of the most important things I can communicate to you about autism history is that, for a very long time, the autism world has trailed the broader developmental disability world by at least one, sometimes two or three, decades," Ari Ne'eman, the cofounder of the Autistic Self Advocacy Network (ASAN), said in an interview. "And so, things that you see in the broader developmental disability [advocacy] that become unacceptable are still acceptable in the autism world."

Ne'eman first came to prominence when he and a nascent ASAN protested ads by New York University's Child Study Center that portrayed autism as kidnapping children and sending ominous messages saying, "We have your son. We will make sure he will no longer be able to care for himself or interact socially as long as he lives." Ne'eman said the ads were reminiscent of the old telethons hosted by Jerry Lewis to raise money for the Muscular Dystrophy Association. Disability rights activists had spent years protesting these telethons, which were demeaning and treated disabled people as objects of pity, and many organizations ditched the toxic idea. But autism groups continued to use that same fear-and-pity archetype to raise money.

"Similarly," Ne'eman said, "you had a greater willingness to support institutionalization, segregation, and even what we would consider abusive treatments," such as shock therapy, as well as bogus cures, like secretin injections and chelation therapy, which involves injecting a binding agent into the bloodstream to remove toxic chemicals.

The problem with focusing on the parents of autistic people instead of the autistic people themselves is that when these two sides clash, society tends to sympathize with parents. A 2011 survey published in the *Journal of Children and Media* examined coverage of autism in publications read largely by mothers (who often wind up as the primary caregivers for disabled children), such as *Parenting, Working Mother,* and *O: The Oprah Magazine,* compared to ones with mixed readership, like the *Atlantic Monthly,* the *New York Times, Reader's Digest, Maclean's, Time,* and *Newsweek.* The publications targeting mothers covered autism like it was a dire affliction. One 2006 article from *Parents* magazine quoted a mother who called her son's autism diagnosis "a gun pressed against the side of my head" and said she thought "the world had just ended." Others talked about moving "heaven and earth" for their children, with one article saying parents "have to be creative—and persistent—about getting the help they need."

The problem with these narratives is that they paint autism as tragic, which it's not. It's a disability the same as any other, and while it requires work, it should not be described as the end of the world; to do so is to compare your child to the apocalypse. Furthermore, if autism is a tragedy, then some parents feel entirely justified in doing whatever it takes to "cure" their kids of this malady, including forcing them to drink bleach or putting them on bunk diets, harming or even killing their children in the process.

Sadly, extreme stories like those are all too common, yet frequently, parents who murder their autistic children are given leniency in both the eyes of the law and public opinion. In 2013, fourteen-year-old Alex Spourdalakis was drugged and stabbed repeatedly by his mother, Dorothy, and his godmother, Agatha Skrodzka. But when introducing a news segment about the murder on *CBS This Morning,* Gayle King said, "The case is extreme, but it shines a light on the struggles of hundreds of thousands of families coping with autism." Similarly, Polly Tommey, who produced a documentary about Spourdalakis's murder, said, "Dorothy was like any other autism mother, desperate to get help for her child" and that "his death didn't need to be. It was because there wasn't anything in place for him." The local NBC affiliate said the two killed the boy "to end the teen's suffering from autism." Three years after the killing, Spourdalakis's mother and godmother pleaded guilty to involuntary manslaughter and were released from prison days later.

While plenty of autistic people and their loved ones do face a labyrinth of obstacles, there is never justification for taking another person's life simply out of desperation or exhaustion from fighting for services. Furthermore, there are countless parents who have children who require significant support and who do not murder their offspring, and stories like Spourdalakis's is an insult to every one of those parents. And all this is to say nothing of how it devalues the lives of autistic people. Autistic people do not need someone to decide if their lives are worth living; they need loving parents who will not inflict harm on them no matter what.

A parent-centered focus on autism has also allowed parents to justify subjecting their children to other heinous treatment. The Judge Rotenberg Educational Center (JRC) in Massachusetts is a day and residential school for students with developmental disabilities that used graduated electronic decelerator (GED) to administer shocks when students exhibit harmful behavior. In a report on the subject, Juan Méndez, special rapporteur on torture for the United Nations, wrote that the shocks violated the UN's conventions against torture. Matthew Israel, the center's founder, was forced to resign in 2011 and agreed to five years of probation. In 2020, the Food and Drug Administration banned the shocks, saying they "present an unreasonable and substantial risk of illness or injury to the public."

But for years, the JRC's biggest defenders were parents. A 2008 profile of the JRC in *Boston* magazine concluded by featuring a hearing with a parent named Eddie Sanchez, the father of a son who received treatment at the center. At one point, Eddie took his son Brandon off the machine, which led to Brandon slapping himself and twisting his neck hard against his shoulder. Eddie Sanchez said at the hearing, "If it were not for this program, my son would be dead. And if it weren't for that, I would blow his freakin' brains out. That's what I would do for my son."

This comment is disturbing. It implies that if Brandon were not undergoing a draconian treatment that uses shocks, Eddie would murder his own son, manipulatively saying he would do it *for* him, as if to take him out of his misery. But instead of saying so, the *Boston* magazine reporter concluded the article with "This is love. It is so strong that the parents of the Judge Rotenberg Center would rather shock their children than see them hurt any more, so strong that they would rather kill their own than have them face life without the machine."

None of this is to say that political autism advocacy does or should pit parents against autistic self-advocates. Many times, the two groups are important allies for effecting change. As you will see in this book, parents often use their advantages as neurotypical people to better advocate for their kids' rights. The main distinction is that parents must recognize when their own desires come at the expense of autistic people's well-being.

The Anti-Vaccine Movement Infects Politics

One of the other bogus theories Bernard Rimland praised was that of British physician Andrew Wakefield, who in 1998 published a study in the *Lancet* that implied a connection between autism and the measles-mumps-rubella vaccine. It was later revealed that Wakefield had not disclosed that he been paid by attorneys representing parents of children who had filed lawsuits against companies that made vaccines. Wakefield lost his medical license in 2010 and the *Lancet* retracted the study, but the damage was already done. In 2000, after the publication of his initial study, Wakefield found an audience in the United States and in the halls of Congress when he testified before the House Government Reform and Oversight Committee. Before then, the committee's chairman, Republican Dan Burton, was perhaps best known for espousing another conspiracy theory: that President Bill Clinton's aide Vincent Foster did not kill himself but was murdered. Burton went so far as to reenact the "murder" by shooting a cantaloupe as a stand-in for Foster's head to prove his case (some reports stated that Burton had shot a watermelon, rather than a cantaloupe, which earned him the nickname of "Watermelon Dan").

In the 2000 hearing regarding vaccines, Burton spoke anecdotally about his granddaughter, who he said nearly died after she received the hepatitis B shot, and his grandson, who he said became autistic after he was vaccinated. Burton said that there was emerging evidence about a connection between vaccinations and autism in "some children" and that "we can't close our eyes to it."

Even during that early hearing, Wakefield's theory was criticized. Representative Henry Waxman of California, the ranking Democrat on the committee, warned that this could cause a dangerous drop in immunization rates. "I think those hearings did a lot of disservice because, to a large number of people, that underscored the false notion that their children could be protected from autism simply by not getting vaccinated for diseases that could be prevented," Waxman told me in 2019, citing the measles outbreaks that had happened that year. "It could have been completely prevented had people not been afraid to vaccinate their kids."

It wasn't all just Republicans either. In an August 2009 hearing on autism, Democratic Senator Tom Harkin of Iowa grilled Thomas Insel, then the director of the National Institute of Mental Health, about vaccines and their link to autism. In response, Insel said, "There is no evidence at this point of any association between vaccines, the number of vaccines, the kind of vaccines or the increase and vulnerability to autism."

What was especially shocking about Harkin asking these questions was that he was the chief sponsor of the Americans with Disabilities Act. Harkin's history as a disability advocate—he delivered a floor speech in sign language upon the act's passage in 1990—implies that he was not acting maliciously. Rather, as a policymaker, he was echoing his constituents who had asked him about vaccines.

"I went through a period where I talked to so many people who had children with autism who became convinced that it was caused by vaccines," Harkin told me. "So, I said it's worth a study. It's worth looking at. I still feel that there's no real evidence that vaccines have anything to do with autism. However, I still feel that the regimen for vaccines for babies is still too ambitious."

Even though Harkin is walking back on the idea that vaccines cause autism, he still parroted the talking point that spacing out vaccines would be better (in fact, falling behind the vaccine schedule could leave children more vulnerable to diseases). Unfortunately, the tireless efforts of scientists, self-advocates, and public health officials still have not quieted the continued mistrust of vaccines. Furthermore, the fact that many people see vaccinations as a conspiracy means they are unconvinced by logic and the reasoning of experts. They perceive the autism/vaccine debate as an attempt to silence them for telling the truth.

The vaccine debate also divided many within Autism Speaks. In 2007, Katie Wright, the mother of autistic son Christian, placed herself on the side of "the Mercurys," a faction named for their belief that mercury in vaccines is a cause of autism. Her parents and the organization's founders, Bob and Suzanne Wright, put out a statement saying "Katie Wright is not a spokesperson" and that "her personal views differ from ours." Alison Singer, who was the Autism Speaks chief executive, quit the organization in 2009 because she felt it focused too much money on the link between autism and vaccines. Singer would later go on to start the Autism Science Foundation.

Vaccines came to dominate much of the political discourse about autism on the presidential campaign trail too. In 2008, Republican presidential nominee Senator John McCain and Democratic nominee Senator Barack Obama spoke about the spike in autism diagnoses and a potential connection to vaccines.

"We've seen just a skyrocketing autism rate. Some people are suspicious that it's connected to the vaccines, this person included," Obama said in reference to a person in the crowd at a rally in Pennsylvania. "The science right now is inconclusive, but we have to research it." At a rally in Texas, Obama's Republican opponent, McCain, said that it was "indisputable" that autism was "on the rise" among children. "And we go back and forth and there's strong evidence that indicates it's got to do with a preservative in vaccines."

While Obama dropped the language about vaccines, his rhetoric at times perpetuated faulty ideas about autism. In 2013, he said, "We're still unable to cure diseases like Alzheimer's or autism, or fully reverse the effects of a stroke."

It's unfortunate that the debate about vaccines gained so much steam in the twenty-first century, because it is completely divorced from the needs of autistic people. These politicians and advocates were not helping autistic young adults who wanted to go to college. They ignored the community's difficulties with unemployment and the question of whether autistic people should be paid subminimum wage. Anti-vaccine quackery infected our politics—ironically—and wrought havoc on America's body politic. And like any other illness, the group with the least amount of protections—in this case, autistic people, who had few allies and little political capital—was the hardest hit.

The Self-Advocacy Movement

While politicians bickered over vaccines, and parents dominated the advocacy movement, a progressive undercurrent was slowly rising to the surface.

In 1993, Jim Sinclair, one of the cofounders of Autism Network International, delivered an address titled "Don't Mourn for Us." Sinclair, who identifies as intersex, delivered the speech in response to hearing parents talk about grieving for their autistic children. Sinclair argued that the parents' grief did not stem from autism itself but rather the loss of the child they'd hoped to have.

In this same speech, Sinclair introduced a radical idea: "Autism is a way of being. It is not possible to separate the person from the autism," they said. "When parents say, 'I wish my child did not have autism,' what they're really saying is 'I wish the autistic child I have did not exist.'"

More than a decade before the Combating Autism Act and George W. Bush's search for a cure, Sinclair pointed out the hurt caused by this rhetoric. They said that when autistic people hear this, they hear "that your greatest wish is that one day we will cease to be, and strangers you can love will move in behind our faces."

Sinclair's words paved the way for shifting the onus of change away from autistic individuals and onto society as a whole. "Yes, there is tragedy that comes with autism: not because of what we are, but because of the things that happen to us," they said. "Be sad about that, if you want to be sad about something. Better than being sad about it, though, get mad about it—and then do something about it."

Sinclair's lecture served as a spark for the neurodiversity movement, the concept that autism and other disabilities, like dyslexia, dyspraxia, ADHD, and so on, are normal variations in the human population and do not require a cure but rather accommodation and acceptance. Sinclair also started Autreat, a getaway for autistic people where they could be themselves. The work of Sinclair and others, like activist and artist Donna Williams, whose book *Nobody Nowhere* was one of the first memoirs by an autistic adult, and Xenia Grant, laid the groundwork for future advocates.

The fear of becoming a "self-narrating zoo exhibit," as Sinclair put it, is one reason I did not want this book to become a memoir. I do not fault other autistic people's decisions to write memoirs. They were integral to getting nonautistic audiences to understand that autistic people had agency, and those works have had a great influence on me. But when I decided to start writing about autism as an autistic person, I chose to place myself within the larger context of autism's narrative. I know plenty of parents clamor for stories of autistic people like me as inspiration, but the truth is, I'm not special. I want them to see my story as a projector that broadcasts the stories of other autistic people. If I am going to write about myself, I am taking as many autistic people as I can along with me.

Ari Ne'eman, the cofounder of the Autism Self Advocacy Network, isn't much interested in putting his narrative on display either. The first time I interviewed him—in 2015, for my piece for *National Journal*—he informed me he was twelve when he was diagnosed with Asperger's syndrome, then he went on to talk about policy. Nowadays, Ne'eman is well known as an outspoken advocate for the rights of autistic people and disabled people writ large. But even before he founded ASAN in 2006 with Scott Robertson, he was reaching out to groups like Sinclair's ANI and Aspies for Freedom—an organization that had petitioned the United Nations to recognize autistic people as a minority group—to learn how to affect policy outcomes.

"The way I saw the work we were doing in founding ASAN was building a sword and shield that could take these concepts that had been incubated by Jim Sinclair and [Autism Network International] and expand their reach and use them to defend our community against trends that were very alarming to many of us," he told me. Similarly, ASAN adopted the disability rights mantra "Nothing about us without us" as its slogan, firmly declaring its goals in terms of disability rights.

The disability rights lens was apparent in Ne'eman's and ASAN's opposition to NYU's ad campaign. Ne'eman successfully coordinated with prominent disability rights activists and created a media pressure campaign to get NYU to pull the ads. ASAN, while still much smaller and less funded than groups like Autism Speaks, would grow in influence over time. In addition, it launched its Autism Campus Inclusion Leadership Academy, which helps autistic college students better advocate for themselves on their own college campuses. Some of the people included in this book are alumni of the program.

Eventually, Ne'eman was appointed by President Barack Obama to serve on the National Council on Disability and was a public member of the Interagency Autism Coordinating Committee, which became a federal advisory committee under the initial Combating Autism Act. The Combating Autism Act was reauthorized in 2011, but when it was up for reauthorization again in 2014, autistic self-advocates began an aggressive campaign to rename the law, using the Twitter hashtag #StopCombatingMe. The legislation was reauthorized as the Autism CARES (Collaboration, Accountability, Research, Education, and Support) Act.

Thanks to the work of advocates like Ne'eman, the culture around autism began changing, and in turn, the political rhetoric changed as well. During his last year as president, Obama guest-edited an issue of *Wired* magazine. In an interview, Obama was asked about autistic researcher Temple Grandin's theory that Mozart, Nicola Tesla, and Albert Einstein were autistic. Obama responded by saying, "They might be on the spectrum," showing he understood the vernacular of autism.

In that same interview, when he was asked about what would happen to society if autism was eliminated, he said, "That goes to the larger issue that we wrestle with all the time around [artificial intelligence]. Part of what makes us human are [*sic*] the kinks. They're the mutations, the outliers, the flaws that create art or the new invention, right? We have to assume that if a system is perfect, then it's static. And part of what makes us who we are, and part of what makes us alive, is that we're dynamic and we're surprised."

Of course, Obama was succeeded by Donald Trump, who has

a fraught history with autism. Bob Wright, the founder of Autism Speaks, tweeted his support of Trump during the 2016 campaign. But their relationship goes back years. When Wright ran GE Capital, the financial services division of General Electric, Trump, then at the peak of his business-development game, tried to persuade Wright to move the offices of RCA and NBC, which GE had just acquired, to one of a planned series of towers he was going to call Television City, but the pushback scared away NBC. Wright was also head of NBC/Universal when Trump's reality television show *The Apprentice* first aired, in 2004.

Trump has been generally friendly toward Wright's ideas about autism (Wright stepped down as Autism Speaks chairman in 2015). Wright wrote in his memoir that his wife was "bitterly disappointed" that Obama failed to light the White House blue, which was then the color of the organization's logo, on World Autism Awareness Day. Trump did so his first year and in his proclamation for Autism Awareness Day on April 2, 2017, he said his administration "is committed to promoting greater knowledge of ASDs and encouraging innovation that will lead to new treatments and cures for autism."

Trump's decision to do this angered many autistic self-advocates and further showed the divide between them and the Autism Speaks organization. For the nonprofit's first ten years, none of its board members were openly autistic, and at the same time, many autistic people vocalized that they did not want to be cured. Similarly, in 2009, the organization put out an ad directed by Academy Award–winning director Alfonso Cuarón titled "I Am Autism," which depicted autism as a menacing force.

"I know where you live and guess what? I live there too," the voiceover said, adding that it worked faster than deadly diseases like pediatric AIDS, cancer, and diabetes combined. "And if you are happily married, I will make sure that your marriage fails," the voice went on, pledging to bankrupt families (there is some irony, of course, that a millionaire executive's charity would put out such an ad). The ad ultimately faced massive pushback, and it was removed from its website.

In 2013, Suzanne Wright posted a piece on the nonprofit's website criticizing the lack of action on autism, saying, "We've for the most part lost touch with three million American children, and as a nation we've done nothing" and lamenting, "These families are not living. They are existing." Once again, the narrative turned to mourning families of autistic people. It wasn't autism that caused parents to "lose touch with their children"; it was a lack of understanding about autism. Yes, there are difficulties accessing services for autistic people, but to portray autism as a tragedy is to demean the value and lives of autistic people and turn them into burdens. The op-ed was so offensive to John Elder Robison, the autistic author of *Look Me in the Eye,* that he resigned from the Autism Speaks treatment and advisory boards.

"That's a tagline for fundraising," Robison said of that tragedy-centric tactic back in 2015. "But I think that, if you're leading an organization that represents autistic people and you say things like that, you have to recognize the powerfully corrosive effect that will have on the psyches of people who are themselves autistic."

Separate from Trump's connections to Autism Speaks, during the 2016 campaign, Trump reportedly met with Andrew Wakefield, the discredited author of the infamous vaccine study.

Ironically, Autism Speaks removed language about a "cure" from its mission statement in 2016, saying on its website that "research found that there is no single 'autism,'" and that science revealed that "there will be no single 'cure.'"

The year before, in 2015, Autism Speaks announced the first two autistic members appointed to its board of directors, Stephen Shore and Valerie Paradiz, who is now a vice president of the organization. In 2020, Autism Speaks replaced its blue puzzle-piece logo with a multicolored piece, which was said to reflect the variety within the autism spectrum. But for many autistic people, those actions came too late and did not undo the damage done.

Since the rise of autistic advocates, many nonautistic advocates have contested their influence, saying that their resistance to a cure ignores autistic people who have higher support needs. Jonathan Shestak, the cofounder of Cure Autism Now (which would later be folded into Autism Speaks), criticized Ne'eman in the *New York Times,* saying, "He doesn't seem to represent, understand or have great sympathy for all the people who are truly, deeply affected in a way that he isn't." Wright wrote in his memoir that "high-functioning autistic adults" do not "want to be associated with the autism spectrum." Wright added that many of them "are satisfied being who they are."

But this is a straw man, since many autistic people like me who have been considered high-functioning recognize that functioning labels do not always accurately describe them. Someone who on the surface might appear high-functioning can have difficulties that are different from those who require more supports. It is because we as autistic people realize that we have much in common with those who cannot speak or who need more services that we want to make the world more adaptable for them too.

Lydia X. Z. Brown, a lawyer and autistic activist, said in 2018 that anyone who is on the "right of the middle on the opposite end" of the spectrum is "closer to anyone who's on the far autistic end than anyone on the neurotypical end."

I know I can pass through this world without being detected as autistic. I am sure that many people will question whether I can accurately tell the stories of autistic people who are different from me. And the truth is, I couldn't if I didn't include their voices, which I am trying to do here. But in the same respect, the more I have learned about autism, the more I've realized that the differences between those of us who can go undetected and the more visibly autistic people are smaller than one would believe.

Furthermore, seeing and reading the work of autistic people with higher support needs helped me come to accept my own autism. The first time I saw an autistic person on the news was in 2007, when CNN medical correspondent Dr. Sanjay Gupta profiled Mel Baggs, who had posted a video on YouTube called *In My Language.* Baggs couldn't speak but used their body to communicate, and seeing this, I felt chills. I felt as if I had been an immigrant in a foreign country my whole life and I'd just heard someone speak my native tongue. There might be different dialects, but we had similar grammar. I remember my mom asking to change the channel because it brought back bad memories of me being bullied, but I couldn't stop watching.

Functioning labels only go so far and don't actually describe what autistic people can do—for example, whether they can speak and hold a job or not. Hari Srinivasan, one of the first nonspeaking autistic people admitted to the University of California, Berkeley, said the functioning labels are stigmatizing.

And many autistic self-advocates have worked to incorporate autistic people with intellectual disabilities into the movement. For example, as you will see later in this book, as companies have sought to hire more autistic people who need fewer supports, autistic self-advocates are working to ensure those with higher support needs get the same opportunities.

The movement to trust autistic self-advocates is one that is finally beginning to bear fruit. The Pennsylvania state legislative hopeful Jessica Benham got into autism advocacy when she met Cori Frazier, who was running the Pittsburgh wing of ASAN. After coming to meetings and getting involved, Benham started staying around the group in the basement of their local human-services building. As ASAN was moving away from a model that had local chapters, Frazier, Benham, and others decided to start the Pittsburgh Center for Autistic Advocacy.

"[It] does everything from peer support to political advocacy to training both for self-advocates and for organizations," Benham told me. "And it's been really exciting to be part of that and to watch that grow and develop into something that's really built a space for the autistic community here in Pittsburgh."

Outlets that focus on autistic people's experiences laid the groundwork for Benham seeking office; they give autistic people agency and a platform to change policy that inevitably encourages some to run for office themselves.

And just as their advocacy will look different, autistic people campaign differently than their neurotypical peers. Benham recognizes that on the campaign trail, there are pluses and drawbacks to being autistic. She knows that campaigning as it exists now is not accessible. At times, she deals with the same executive-functioning difficulties many autistic people have, like forgetting to eat, which is when having a staff is helpful.

On that February day in Pittsburgh, as Benham and I exit her car and arrive back in her office, she tells me that making history as the first autistic woman elected to a state legislator is not the goal of her campaign.

Later, as we are wrapping up the interview, I ask Benham for some general closing remarks, and she talks about why it's important to have a legislator who is fighting for labor, infrastructure, and all the other things she supported as a community organizer before talking about being autistic.

"So, you know it's certainly a milestone to have an openly elected autistic individual to a state legislature, but for me, that . . . it's only a small part," she said.

Benham again is trying to broaden her appeal, which is understandable. It is the same thing many politicians who are part of a minority group do, an attempt to emphasize what they have in common with their constituents. Even if they require extra assistance, autistic people are entitled to campaign the same way as other politicians.

For all my excitement at the feeling that Benham could be knocking down barriers, I've also recognize a soothing sameness throughout Benham's and my time together. I've followed candidates campaigning in the past, been to countless political-party and interest-group functions, and seen plenty of politicians press the flesh while I stand on the side with my notepad or sit at a rickety table with my laptop. There is something relieving to me that Benham and I are both members of this Spectrum Generation and we have the opportunity to be judged by our merits, not our autism. People can choose to vote for Benham or her opponent. Readers can think I am an honest broker or a fake-news hack. Either way, we want to be measured by the same metrics every other politician or journalist is measured by.

Later that day, as I am heading to the airport to catch a flight back to Washington, I check the Steel City Stonewall Democrats' endorsement results on Twitter. Benham has won their endorsement.

A few months later, she won her election.

2

"In My Mind, I'm Going to Carolina"

EDUCATION

Huntington, West Virginia, is near the border of Ohio and Kentucky. To get there the Sunday after Thanksgiving in 2018, I flew from Washington to Charlotte and then to the Tri-State Airport. My ride from the airport was filled with scenic views of mountains that kiss the sky. Weeks after the 2018 midterm elections, I joked with one of my Lyft drivers that I was the only political reporter who was not there to cover the election. And I certainly wasn't there to write one of those god-awful "Trump Country" profiles about "forgotten voters" in Appalachia.

National political reporters who cover West Virginia usually write about the decline of the coal industry, the gripping poverty, or the opioid epidemic that has ravaged the state (at one point leading to 41.5 deaths per 100,000 people). These were usually the rationales given for why the state had shifted from voting for Democrats in the twentieth century to voting overwhelmingly for Donald Trump in 2016. Hence I was puzzled when I was interviewing someone at the University of Alabama about its autism students' program, and Megan

Davis, the director of UA-ACTS, mentioned they had modeled their program after Marshall University's in West Virginia.

I later learned that this program was one of the first in the country, and West Virginia's leading role in autism history was largely thanks to Ruth Christ Sullivan, the cofounder of the National Society for Autistic Children (later the Autism Society of America). While Bernard Rimland was largely focused on searching for a cure for autism through various remedies, Sullivan was focused on fighting for services for children like her son Joe, who would serve as one of three inspirations for Raymond Babbitt in the movie *Rain Man.* When Sullivan relocated to Huntington, she started NSAC's Information and Referral Service, which helped parents find resources for their kids, out of her home. West Virginia was also the first state to specifically include autism in its special education laws.

In 1984, Sullivan led efforts to create the West Virginia Autism Training Center at Marshall University to serve families of autistic people throughout the state. In 2002, it began offering services for autistic students through its College Program for Students with Autism Spectrum Disorder.

The program office is located on the third floor of Marshall's Old Main Suite building, the oldest building on the university's campus. I had organized a trip there and the program's team agreed to set me up with a handful of students to interview about their experiences as the semester was closing. While there, I met Richie Combs, who was in his final year of college studying environmental science. Combs told me that he benefited from receiving extra time on his tests and having advisers at the center speak to professors on his behalf when he was overwhelmed. Zachariah Lewis, who was graduating that December with a major in history and a minor in psychology, said one of the best things about the center was how it gave him the ability for him to be himself. "It encourages a sense of solidarity [and] camaraderie," he said, noting that he was able to engage with people from all different parts of the country—and the world—and make lasting friendships.

These students were more accepting of their autism than I was at their age. The changes that have occurred in both policy and diagnostic criteria mean that college has become a possibility for countless autistic people who were previously thought incapable of matriculating. Though, in truth, plenty of autistic people have gone to college; unfortunately, many did not realize they were autistic or they were misdiagnosed, and as a result, they struggled to get through school. Others were seen as incapable of any type of adult life, so college was not seen as an option. Nowadays, many autistic students still deal with a lack of accommodations, internalized ableism, and difficulty navigating the social aspects of college.

Academic Accommodations

In my senior year of high school, I was living with my mom and my sister in California. My mom proposed I attend a community college before going to a university to test my independence. I think she was concerned that I would have difficulty with executive functioning and carrying out independent tasks while balancing a full academic workload. Going to a community college could be a bridge between the more routine schedule of high school and being thrown straight into college life. I was reluctant because I saw community college as "high school with ashtrays."

I wanted to go to a local university, California Polytechnic University, which was where my guitar teacher JohnPaul Trotter had gone; I wanted to get a degree that would help me become a journalist and then find a way to write about either music or politics. Part of me still aspired to be a professional musician, and journalism was a contingency plan. But I was terrible at managing my time and always procrastinated until the last minute for nearly everything. It finally came back to bite me when I missed the application deadline to apply for California public universities because I put off asking my mom what my Social Security number was and then, after I finally worked up the courage to do that, I was too frozen with anxiety to ask her for the application fee. Thus, I had no option but to enroll at Chaffey College.

Looking back, I now know that community college should not be a source of shame. In truth, I learned and grew a lot at Chaffey. Throughout high school, I was relatively smart but had little work ethic and thought I was a bit above everyone else, despite the fact that I never took AP classes and regularly struggled with math. I could score As and Bs pretty easily in other subjects because I could absorb things quickly. At Chaffey, I gained much-needed discipline. I went to the tutoring center daily to work, developed smart study habits, learned to ask my professors for help, and prioritized studying.

And it paid off. My Middle East history professor Tim Greene inspired me to want to learn Arabic, which later ignited my temporary desire to be a war correspondent (much to my mom's horror). Plenty of the professors there were attentive to students' needs and held frequent office hours. My fellow classmates were people coming back from military service, people who'd gotten hit hard by the economic downturn of 2008, and moms whose kids had grown up so they finally had the chance to get an education of their own. They were and are good people, and community colleges are a vital lifeline for many.

Autistic students are about twice as likely to enroll in a community college than in a university, according to a May 2014 study in the *Journal of Autism and Developmental Disorders,* which found that 81 percent of college students with ASD were enrolled in community college at some point after secondary school. But the authors also found that students who studied science, technology, engineering, and mathematics (STEM) were twice as likely to transfer to a four-year university than those who studied non-STEM subjects. At the time, I didn't know that, but it is easy to feel that autistic people who don't want to pursue these subjects aren't treated with the same priority. It is good that autistic people have a means to pursue a bachelor's degree, but this dichotomy abandons far too many autistic people.

Another huge benefit of community college for me was that it gave me time to shed my dislike of asking for disability accommodations. In the classroom, some common accommodations are extra time to take tests and isolated areas in which to take them so the student is not too overwhelmed by surroundings. I loathe the term *special needs* because it implies that disabled people are getting preferential treatment rather than the assistance that is their right. At Chaffey, I initially felt guilty asking for accommodations, since the mere fact that I was a student made me question whether I deserved accommodation services in the first place. Plenty of disabled people don't make it to college, so I felt I must not be truly disabled if I was there. I also worried that people would think I was "taking advantage of the system" or put an asterisk next to my success.

Swallowing my pride was a major source of tension between me and my mom when I was at Chaffey. I eventually relented when my mom visited the school's offices to ensure I got services. She insisted that they were there for a reason: to make sure disabled students got fair treatment. I ended up getting extra time on my math tests in a separate, quiet space. And thanks to this accommodation, I eventually did well enough to get into the honors program at Chaffey, which allowed me to take accelerated classes and put me on a fast track to a four-year university.

Ironically, when I called to enroll in the honors program, they said they didn't know how to handle my disability services because they couldn't recall ever having a student with accommodations in the program before. So often, if disabled people somehow succeed, either our disabilities or our accommodations are questioned.

Timotheus Gordon also had a complicated experience with accommodations. He told me that in high school, he felt that they were more of an impediment than a service. But when Gordon was at the Savannah College of Art and Design earning his master's degree, he struggled with writing graduate-level essays. He finally met with his teacher when he was at the point of failing a class in his first semester. That teacher suggested he get more time on tests and get counseling services, an offer that Gordon at first rebuffed. He did not want his education to "feel remedial," he said, adding that he'd already been through that as a child. But now his success was on the line.

"I didn't want to fail," Gordon said. "So that's when I finally swallowed my pride and got some time on my tests."

M. Remi Yergeau, an associate professor of English at the University of Michigan, was diagnosed as autistic while in college and told me they had a similar experience with accommodations. "There's this element of disbelief or a tendency to see students as whiners, as being overly sensitive, as not wanting to deal with hard problems," they said. "It's rage-making . . . for a lot of disability activists and academics who are disabled or do disability studies."

This general distrust makes the few professors who understand disability accommodations extremely refreshing. When I first disclosed to my cinema professor at Chaffey that I was autistic, he was incredibly empathetic. He never specifically singled me out in the class or made things easier on me, but it led me to trust him and I wasn't afraid to go up to him when I had a question about classwork.

Community college is a stepping-stone for many autistic students on their way to a four-year university, myself included. Hari Srinivasan, who was one of the first nonspeaking autistic people to be admitted to the University of California, Berkeley, attended San Jose City College.

"The smaller nature of a community college meant that my counselor knew most of the instructors, and would steer me towards courses that had more understanding instructors which makes a huge difference in your college experience," Srinivasan told me in an e-mail. "This is not something you can ask for in writing."

When I was enrolled in the honors program at Chaffey, I started looking at colleges with good journalism programs and saw that two of the best public universities for journalism were the University of Florida and the University of North Carolina at Chapel Hill. I hated humidity, so I figured it would be better to go to North Carolina (I didn't know that North Carolina also got humid in the summer). After I got in and learned I'd received enough financial aid, my dad helped me move to Chapel Hill. We met with my academic adviser and found all the important routes to get to class but visiting disability services was not on our agenda. Maybe we forgot, or maybe we thought it was unnecessary. In hindsight, I'd postulate a third reason: I felt I had made it as a college student. I saw my autism as something I had conquered and overcome, now that I was enrolled at a prestigious four-year university. It would take time and a few close calls before I accepted the truth—that autism is part of who I am.

Much like the anxieties I had at Chaffey, I felt that if I used special accommodations at UNC, I was somehow cheating. I saw people in my classes busting their asses in the library, and I feared I would be taking a shortcut if I got help that I felt I didn't "really need."

My assumptions would be shattered roughly two months into my time at UNC during an exam for my African music class when all the answers somehow evaporated from my memory. In an act of desperation, I walked up to my professor's office after the test. Shaking, I told him that I was autistic. I don't remember what I said verbatim, but I explained I sometimes had trouble comprehending abstract ideas and taking tests in "normal" academic settings where I was timed. I feared he would think I was making up an excuse for doing poorly. Instead, he immediately called the student learning center and set me up with Dr. Theresa Maitland, a guidance counselor. I eventually passed his class with a B, and while I never took another class of his and I think I saw him only one more time on campus, I love him for making that call.

Dr. Maitland—or "Doc," as I would come to call her—immediately gave me strategies to study better. I never really used other accommodations, but her office would be both my sanctuary and my war room. She became my confidante, my adviser, and my surrogate grandparent. She never made me feel like I was flawed or weak. She encouraged me to exercise, pushed me to find extracurricular activities, and helped me navigate interpersonal relationships. She told me that it was okay to go see professors during office hours because they were there for a reason.

But as with all instances of kismet, I can't help but wonder what the alternative could have looked like. What if I hadn't found my music professor in his office that day after I bombed the test, and he hadn't set me up with a guidance counselor? Or, worse, what if he'd interpreted my confession as a lie, an attempt to skate along in his class? What if he had called the disability services office and they were already closed for the day? Or what if they'd refused to see me because I had not filled out the proper paperwork at the beginning of the year? These what-ifs plague me.

M. Remi Yergeau, the Michigan professor, told me that this is the main reason autistic people fear asking for accommodations—it feels as if we're seeking out special treatment and that we're the problem. But asking for help is about addressing structural problems that hold autistic—and all disabled—people back.

"It cannot end there, because that would be an admission that, yes, disability is a personal problem, rather than seeing how disability is created by inaccessible environments, how disability is created through discrimination and bias about what represents normal, what represents a normative body or mind," they said.

Hari Srinivasan said Berkeley's disabled students program is incredibly accommodating, but the difficulty comes with what to ask for, since it is hard to accommodate fluctuating sensory issues. On some days, he needs a minimal amount of support, but on other days, he can be overwhelmed with absolute sensory overload, which leads to him feeling ready to fall apart. "So, my already limited communication can shut down and I won't even be able to ask for help," he explained. Srinivasan said few people understand that each person's nervous system reacts differently to various stimuli, and this can change by the hour.

Until the world better understands the true needs of disabled people, Srinivasan said, all autistic people can do is ask for the standard accommodations like extra testing time; an aide for communication, regulation, and behavior support; permission to use an iPad or laptop in exams and essay responses; and alternative media for textbooks.

But Srinivasan admits, "This imperfect solution is somewhat stressful."

Still, the fact there are services and attempts to make schools more accessible is leaps and bounds ahead of what was available for many autistic people in previous generations. One woman I met online who goes by Aria on social media told me college was not the right fit for her. Aria, who was undiagnosed until she was twenty-nine, initially attended Grove City College (a "conservative Christian college," according to Google) in Pennsylvania, three thousand miles away from her home in Bremerton, Washington. She went there to study music education with the hope of becoming a band director, but over time, she changed her major to English and then to sociology and psychology.

"Looking back, had I known I was autistic at the time, I would've probably chosen to go to the community college nearby in Bremerton while living at home," Aria told me in an e-mail. At college Aria said she became "horribly depressed and suicidal and self-medicating all the time," so she quit school, fearing she was wasting her parents' money.

Aria, who is now married and has children, seems like someone who would be high-functioning by society's measurements, while Hari Srinivasan is someone the world would perceive as low-functioning because he cannot speak. When he was young, he was even subjected to applied behavior analysis—a type of treatment for autistic people focused on improving communication and social skills. He was not "succeeding" in it, so he was moved to a special education school in middle school. Since then, he's become an accomplished writer at his campus newspaper as well as a poet.

But Srinivasan thrived at college while Aria did not; she was forced to figure things out alone because she did not have a diagnosis and the accompanying accommodations. So while the label *high-functioning* might be used as a compliment (I know people have called me that in the past), it winds up delegitimizing the needs of autistic people who can pass.

Finn Gardiner, an autistic advocate based in Massachusetts, admitted his past experiences using accommodations shaped his reluctance to ask for them. Like Aria, Gardiner dropped out of a small Christian liberal arts college after experiencing what he said were instances of racism and a lack of understanding about disability.

Gardiner, who is both queer and transgender, was also estranged from his parents, which led to struggles with homelessness and poverty. But when he moved to San Francisco, he found a transitional living program for runaway youth that helped with education and employment.

"I had things like check-ins, and they would support students with things like books and registering for classes. That support was both from them and professors who were incredibly dedicated to making sure I got through," Gardiner recalled. After graduating from City College of San Francisco, Gardiner earned his bachelor's degree from Tufts University in Massachusetts and later his master's degree in public policy from Brandeis University's Heller School. He now works at Lurie Institute for Disability Policy at Brandeis.

Having professors and faculty that adequately support students is necessary for any autistic person looking for guidance. For example, there's Lydia Brown, who rose to prominence writing the blog the *Autistic Hoya* while a student at Georgetown University. Brown (whose preferred pronouns are *they* and *them*) taught a class at Tufts University and said they made an effort to ensure the class was as accessible as possible for disabled and autistic students.

"I set very clear and specific deadlines for each of the papers and projects so that students who are very structure-oriented and deadline-oriented would have a clear set of feasible deadlines to complete all the assignments of the class," Brown told me over lunch at a restaurant in Washington, DC. The major difference between Brown's and other classes, though, is that there was no such thing as late work.

"So if you can't finish by this deadline, all that you have to do is tell me when you think you will get it done by, and that's your new deadline," Brown said. "What's worse for many autistic people who make it to higher education is that people who managed to get the higher education tend to be the autistic people that were told 'You're smart and you're good at school,' and then they fall apart in school, and then no one believes you because they're like, 'Well, you're smart. You're not really struggling, you're just being lazy and you're making excuses, you're not trying hard enough.'"

This makes an understanding and empathetic faculty all the more necessary. For me, one such professor was Ferrel Guillory, a man in his sixties with only a few wisps of snow-white hair left on the top of his head who always wore an immaculate dress shirt and reading glasses and who carried traces of an accent from his native Louisiana. Professor Guillory was an elder statesman of political journalism in North Carolina who taught a class on Southern politics. I wanted to impress him, but I tanked my first column for his class, as I'd had trouble grasping the abstract concepts that came with writing a column and could not nail down a specific topic. I felt like a failure.

But a few days later, Professor Guillory invited me to his office, and after closing his door, he asked me, "Do you have Asperger's?" I remember thinking, *What gives you the right to ask me such a personal question?* But instead, I said, "How did you know?" He said he knew people on the spectrum and had decided he would help me whittle down my ideas on a weekly basis to ensure my writing would be a finely sharpened weapon. I felt completely naked and exposed, as if my worst secret had come to light and I could no longer hide, but I also felt relieved. At that point, I didn't care if Professor Guillory's question was appropriate; he was saving my ass, and with his help, I learned how to write better analytical pieces. And God bless him for that.

Gardiner, Srinivasan, and I are all lucky that we had people in our academic careers who were understanding and willing to give us assistance, even if they did not always fully grasp the intricacies of autism. More important, none of us "overcame" our autism or our disabilities. Late disability rights activist Stella Young dubbed these "inspirational" stories about autistic people "inspiration porn" because they are meant to make able-bodied and -minded people feel better about themselves.

The three of us succeeded because there were systems in place that enabled us to succeed. To be inspirational means to be exceptional and somehow extraordinary, but to be successful because of supports and accommodations is to universalize our success. It is to say that other autistic people can do these same things if they have the same resources—thus, we can make the extraordinary just ordinary.

Alternative Approaches

While many colleges and universities are still trying to figure out how to accommodate autistic students using existing disability services frameworks, some are trying to tailor specific programs for autistic people and make an accessible campus. Back at Marshall University, the autism training center provides education for the faculty, staff, and students as well as local employers about autism through its Allies Supporting Autism Spectrum Diversity program. Rebecca Hansen, the director of campus-based services, told me that the program is an hour long and includes a slideshow with an autistic woman student, which helps to counteract the common stereotype that autism occurs only in boys.

"The allies' training has been a really big benefit to helping other people understand the best ways in which to embrace folks on the spectrum so it's not seen as something that can be stigmatizing but more so welcomed and another part of who someone is," Hansen told me in an interview. But Hansen also said getting the proper training for faculty can be a way to better assist autistic people in asking for accommodations when they need them. "We want to eliminate that kind of embarrassment by helping professors understand that these are typical supports."

Hansen said many of the education accommodations requested by autistic students relate to executive functioning.

A lack of executive functioning, in layman's terms, means difficulty in planning and carrying out complex tasks. It's why some autistic people can be extremely intelligent, whether or not they can speak, but struggle immensely with applying knowledge. That sounds like a cheap excuse for laziness. But for many autistic people, it just doesn't come naturally. It wasn't until I learned about executive functioning that I understood why I could memorize all of my notes for class but allow menial tasks like taking out the trash, cleaning my room, and picking up dry-cleaning to go unaddressed for weeks on end. It's not that I liked living messily. I knew I needed to address these things. I just didn't have the natural wherewithal to plan or even to ask for help. This is the case with many of the students at Marshall, Hansen told me.

"A student may struggle with getting to class on time because of a variety of things. It could be anything from a change in the weather pattern; it could be a change in sleep patterns; it could be a challenge with medication routines; it could be that they overslept just like anyone else," she said. Hansen added that was why the program helped faculty realize that "it is not a sign of disrespect" if autistic students miss class or struggle with other tasks.

In addition to offering academic support, the center at Marshall teaches "adaptive living skills" to help students navigate independent living on campus. A skills-building group helps autistic people work on things like time management, developing a "social radar" to interact with classmates, building a reputation on campus, and stress and anger management. All the while, students are fully integrated into the campus. The environmental science major Richie Combs told me that while he usually only needs accommodations in class for extended time on tests, he has also had help with planning.

"I'm usually pretty good at knowing what I have to do. It's just a matter of doing it," Richie told me. The center sends him reminders about upcoming assignments or exams, something that can be helpful for autistic people who have difficulty with time management or setting schedules.

Richie also praised the Discovery Groups at the center, which Marshall's website describes as a program to help autistic students navigate campus; these groups help students with time management, executive functioning, conflict management, and handling stress and anxiety.

But for all of the good that the program does, there might be some parts of the discovery groups that are concerning, and there is also a risk that they could be aimed at changing who the autistic person is. For example, one discovery group is called Developing a Social Radar, and it focuses on undersharing versus oversharing and how both can be either beneficial or detrimental to interpersonal relationships. But while autistic people may benefit from learning interpersonal skills, programs like this put the onus on autistic people to fix their problems rather than on the people around them to be more empathetic.

And while many of these programs can be beneficial, they have a steep price tag. Marshall's, for example, is five thousand dollars per semester as of 2020. Hansen said many students receive assistance for the program through scholarships and West Virginia's Division of Rehabilitation Services. But that assistance is largely dependent on state funding. In 2017, the year before we spoke, Hansen said twenty-three out of the program's fifty students used vocational rehabilitation money.

"They had a major budget cut since then," Hansen said. "So it's harder to get those funds at this point, although it's possible. It's just a matter of justifying it and applying early."

On the other side of the country, Bellevue College, a community college located near Seattle, takes a different approach. Sara Gardner, the director of the Neurodiversity Navigators program, is autistic and has an autistic son. Gardner said it was important that the program not try to change who autistic people are.

"A lot of other programs are trying to turn autistic students into human beings rather than starting from the premise that they already are," Gardner told me. Gardner said the Navigators program focuses on helping students develop an identity, so they learn who they are as autistic people. Gardner said autistic students are often denied this right from kindergarten through high school. Autistic students don't pay any additional fees for the program other than tuition.

Much like Marshall's program, Bellevue's has credit classes that help autistic students navigate school. Some of the topics overlap with Marshall's, like executive functioning and interpersonal communication. But other subjects at Bellevue include navigating college and a career, self-community and advocacy, and occupational wellness. The Bellevue program is modeled on the "social model" of disability, a response to the "medical model" that sees disability as something that needs to be fixed. By contrast, the social model of disability argues that the biggest obstacle to disability is a world around disabled people that does not accommodate them.

For my own part, I know that when I was in school, once I came to accept that I was disabled, it didn't make me want to work any less hard. Instead, it made me aware of my weaknesses and allowed me to plan accordingly. I was willing to go to my professors' office hours for extra help, especially in classes where I struggled, like psychology and computer coding.

Life Outside the Classroom

I felt a sense of accomplishment for surviving my first year at UNC, but I also felt lonely. I was technically a sophomore, but I had moved into the freshman dorms my first year, and as a result, I had little in common with my peers. I was living in a new state where I didn't have any family or friends, which only added to that social isolation.

Loneliness is a common experience for many autistic people. A 2017 study from the *Journal of Autism and Developmental Disorders* found that for most autistic college students, the major problem wasn't the academics but the on-campus aspects of college life. "Over three quarters of the students reported regular struggles with feelings of isolation, being left out, and lacking companionship," the study said. Autistic students had an alarmingly high risk of suicide, with nearly three-quarters of participants saying they "had some form of suicidal behavior over their lifetime," 40 percent of them reporting having made suicidal plans, and 14.6 percent attempting suicide.

To counteract that loneliness, I made the decision to get involved and join a social group. After realizing I did not really belong to either political party and life as a partisan Democrat or Republican was not for me, I decided to return to my first passion, journalism. I signed up to work for the state and national politics desk of the *Daily Tar Heel,* the campus newspaper. It was the fall of 2012, during the presidential campaign between Barack Obama and Mitt Romney, both of whom were making a play for North Carolina. As a test to show I could be unbiased, I readily signed up to cover the North Carolina GOP's election-night party in Raleigh with a few other cub reporters, including Amanda Albright, who would later become a dear friend.

The whole night was exhilarating. In the reception room with a cash bar of some middle-grade hotel in downtown Raleigh, me, my colleague Jacob Rosenberg, and Jeff Kagan swarmed around the hall mining for quotes from Republicans about higher-education policy to feed back to Amanda, who was writing the main story. At one point, one of us broke the news to one candidate (who was already pretty buzzed) that he'd lost his race, to which he replied, "Holy shit, I lost." You could almost see his soul leave his body.

Later on, Jeff and I were tasked with finding Paul Newby, the head of the state supreme court, so we scoured the building until we found him outside his hotel room. We started asking him questions as he was checking his phone for results and he was polite but short with us. I noticed he had an Eagle Scout lapel pin, so I immediately asked him about it and stretched out my left hand (a greeting among former Boy Scouts). Shortly afterward, he invited us into his hotel room, so Jeff and I were in the room with him when he learned he'd won his race. We got the quotes we needed and sent them down to Amanda. That night, despite Obama handily winning a second term, Republicans cleaned up in the state.

As we were walking out, we saw a trash can full of Mitt Romney signs, so I swiped one as a memento of my first election night. Afterward, we went to Cookout and got burgers and Cokes and celebrated the end of our first long election night. As I was going home, my legs almost gave out, and I conked out as soon as I got in bed, but I knew I was going to be covering politics for the rest of my life.

But more than that, I had found a friend group, people who were hungry for stories, followed politics ravenously, and would take whatever ethical measures necessary to get the story, and my love of politics and news made me an asset among them. (Some—like Amanda and Jacob—have had accomplished journalism careers themselves since then.) We could speak a similar language. I now had something to look forward to after my classes and homework (or, honestly, I had an excuse to avoid them), since I could just go to the *DTH,* get a story assignment, and be off and running. Not to mention when I got a good story that was on the front page, I would get praised. I had a home now.

Many of my friends from the paper are still my friends today. Maddy Will, the assistant editor on the politics desk, arrived in DC shortly before me. Dan Wiser, a reserved establishment conservative who was Maddy's superior when I joined, was my roommate one summer when we were interns in DC. My friend Mary Tyler March (yes, she's a journalist and she was named for Mary Tyler Moore) and I have been colleagues on two occasions: once at the *DTH* and another time as editors at the *Hill*. My former colleagues Sarah Brown and Dan Schere and so many others remain incredibly vital to me. The *DTH* gave me these people, and their bonds and ties will endure long after the ink fades and the old print pages turn yellow.

Not Leaving Others Behind

For many autistic people, college is simply not a viable option or there have yet to be environments built for them that are empathetic to their needs. Despite all the strides that have been made toward accommodating autistic people in higher education, for many, the institutional barriers are still too difficult to overcome.

Anyone who follows Andrew Savicki on Twitter under the handle @SenhorRaposa knows he is one of the best political prognosticators in the country. In March of 2018, Savicki wrote that one "hot seat" House race to watch during that year's midterm election was Oklahoma's Fifth District, labeling it a toss-up and saying that Democrat Kendra Horn could win. Major prognosticators all rated the seat as secure for Republicans. But Savicki argued it was "much more competitive than it initially appears." On election night 2018, Horn proved Savicki right. But despite his political wherewithal, Savicki, who was diagnosed in 2015, has been unable to graduate college because of his trouble in math classes. "No matter how much tutoring I get, I can't advance any further until I get those math classes done. So, I'm just stuck in a never-ending loop unable to go forward," Savicki told me.

It is for this reason that many open autistics who work in academia have attempted to make it more accessible to students. In 2015, John Elder Robison, who is the cochair of the neurodiversity working group at the College of William and Mary and himself a high school dropout, urged autistic people working in academia to "come out" as autistic to combat the misperceptions about autism.

"Everyone knows how autism, ADHD, dyslexia and other neurodiverse conditions disable us as children. What we need to balance that are successful adults who attribute their achievements in part to neurodiversity," Robison wrote in a blog post.

Robison noted that whether autistic people in academia were the "least disabled in our community," hardworking, or just the luckiest in our group, they should be serving as examples for parents of autistic people and young autistic college students. I tend to agree with some of that; autistic people aren't any less disabled just because they are in academia, and often, the professor's lectern or the academic adviser's desk can hide a lot of the adversity that comes from ableism in academia. And frequently, as I have noted in this chapter, powers beyond our control—whether that is God, luck, karma, or something else not the focus of this book—plays a large role in autistic success in academia. We do need autistic people in academia to show what autistic people can do when given the right tools. Furthermore, autistic people, including those who require more services, being open in academia show that many autistic people whose disabilities affect every aspect of their lives can *also* summit incredible heights.

"That's the best antidote to talk like 'He's autistic; he'll never go to college,'" Robison wrote. "While it's true that profound disability will leave some of us requiring substantial supports and residential care even as adults, most of us can grow up to live independently and we have great contributions to make."

Just as there needs to be autistic faculty and leadership, there must be neurotypical leadership in academia who prioritize recruiting autistic students. This was the case with Dave Caudel. After graduating high school, Caudel, who did not know he was autistic at the time, joined the army and then held a series of jobs in civilian life before going to college. But when he struggled to advance to graduate school because of poor admissions test scores, Caudel was recruited into the

Fisk-Vanderbilt Master's-to-PhD Bridge Program, which helps underrepresented students get into graduate programs in STEM fields. It was through that experience that Caudel met Keivan Stassun, who founded the bridge program and who has a son on the spectrum.

Eventually, Caudel was approached to lead a program to help autistic people who were having trouble finding employment. Caudel hadn't planned on or prepared for the offer, but he was tempted. "It's something that could make a real difference in people's lives, so of course I had to say 'yes.' This is my tribe. It's weird because I spent most of my life never having that experience," he told me. Stassun and Caudel now lead Vanderbilt's Frist Center for Autism and Innovation, which bills itself as working to promote neurodiverse talent.

Despite initially not thinking he was the right person for the job, Caudel said that he feels a deep sense of obligation to autistic people, particularly ones who faced challenges like he did, such as poverty. Many of his colleagues who have loved ones who are autistic are at an advantage, since they can afford to go to private schools and get special tutors and accommodations. "If we only take care of those wealthy kids who have autism, we're leaving a lot of autism on the table," he said.

Caudel benefited from having a neurotypical ally in that astronomy professor who wanted to help autistic students and who gave him a platform to lead, which is a good example of how to construct an academic environment where autistic people can excel and take control of their education. It's a model in which autistic people are partners instead of merely patients or recipients.

Caudel said he is adamant about making sure that autistic people have a say in the construction of programs, and he reminds his boss that these projects would have more impact if the people working on them and who were paid were autistic themselves. "It would be awesome to say, 'This was built for autistic individuals, by autistic individuals,'" he said. "Not to mention, they're bringing their own brilliance and their own viewpoints."

Initially, Caudel and the center were primarily focused on autistic adults outside Vanderbilt. But he told me that over time, more students approached him and others asking for help around various aspects of college life, like how to speak to an academic adviser or what to do once they graduated. Caudel and his team have worked to change perceptions about autism. He noted one instance in which a counselor told a student he didn't look autistic, so Caudel met with the counselor to help build better understanding.

In response to the students' requests, he helped start the Vanderbilt Autism and Neurodiversity Alliance (VANA), which was open to everyone from students, faculty, and staff to connect as neurodivergent people.

Caudel said he is always conscious of how he can work to add diversity to Vanderbilt's autistic community. "The thing that keeps me up late at night is the fear of which parts of the spectrum am I not paying attention to," he said. "They almost all need help, and every time we hire someone with a slightly different demographic, or slightly different part of the spectrum, it has added value."

One of these hires was Claire Barnett, a graduate of Vanderbilt who wrote about her experiences on campus as an autistic student at the university.

"There are two sides to Vanderbilt's treatment of neurodiversity —cultural and institutional. For the movement to create change at Vanderbilt, it must be addressed from both angles," she wrote for her campus newspaper. In addition, like me, Barnett is a former White House intern; she interned in both the photo office and in Vice President Mike Pence's office during the Trump administration. During the second internship, Barnett reached out to me as a fellow autistic White House intern for advice, and I was taken aback by her intelligence and her blunt nature. Today, Barnett serves as communications director at the Frist Center. But Caudel is quick to note she is good at her job for many reasons, not just because she is autistic. "I remember there were a couple of times when we would meet with some people, and she would tell people, 'I got hired because I'm autistic,' and I said, 'Wait, stop, this is a misconception,'" Caudel said. "[She is] a brilliant communicator. On that strength alone, [she is] perfect for this job."

The community Caudel has built at Vanderbilt makes me think about my final months at UNC, which were some of the most rewarding and the most brutal in my life. By moving to a state where I had never lived and working at the *DTH* and befriending my colleagues there, I found a new home that allowed my world to grow to include so many lovely people. In my final months on campus, I would pass by the quad and be reminded of the days I would ask students questions for stories. On weekends, I would walk past the Dean Smith Center, the stadium named for the revered basketball coach who coached Michael Jordan, and be reminded of going to see the UNC-Duke game with my friends.

I was not oblivious to UNC's flaws. Like most snot-nosed budding journalists, I saw myself as whacking away at the systemic inequalities that kept people out of UNC; my friends and I would regularly rail against the ubiquitous tributes to the Confederacy and segregationists on campus. My love for UNC wasn't unlike my deep patriotism—as a journalist, I've criticized my country's shortcomings out of a desire to see it be what I was taught it should be.

Leaving UNC definitely hurt. Cammie, who had been the managing editor at the *Daily Tar Heel,* asked me to sit next to her at Kenan Stadium during commencement. That Mother's Day in May 2014, I nearly overslept but got up just in time to shower and put on my cap and gown. My mom, Bob, my dad, and Stephanie rushed to Kenan from Raleigh-Durham International Airport. Steph's graduation from

Loyola Marymount University had been the day before mine, so they'd taken a red-eye flight from California to Chapel Hill and arrived in the sweltering heat just as the music started.

I felt the unknown future breathing down my neck, though more realistically, that was probably just the sweat from having to wear a cap and gown in the Carolina humidity. Still, at the time, I had no job lined up and was dreading returning home. I feared if I couldn't make money on my own, all my efforts to graduate would have been for naught. I looked over at my family in the audience, thankful they were there but also scared I would have to go back and lose control of my own future.

But for a brief moment, as the ceremony ended, those anxieties faded as I heard the Clef Hangers, UNC's male a cappella group, sing "Carolina in My Mind." James Taylor wrote the classic about his youth in Chapel Hill (where his father was a professor). For those two minutes, I put my arm around Cammie's shoulder and hummed along as they sang, "And ain't it just like a friend of mine to hit me from behind? Yes, I'm gone to Carolina in my mind," as if UNC and all the people I met there and at Chaffey were hitting me from behind to push me into the real world. Heartbreakingly, I was also leaving the first place I could genuinely call home.

3

"That Ain't Workin'"

WORK

John Marble didn't intend to have a career as an advocate for autistic people in the workforce; in fact, he didn't even intend to admit he was autistic. He spent much of his career working in politics, starting as a White House intern for Vice President Al Gore. Afterward, he did advance work for Gore and then left college to work for Gore when he ran for president. He ultimately served as a presidential appointee to the Office of Personnel Management. It was during that period when Marble was spinning out of control at work, at one point thinking he would go deaf because every sound became white noise (he later learned this was a difficulty with sensory processing). Marble said he saw his autism diagnosis approaching.

"I started seeing the picture of autism, and I fought it, and I resisted it, and I was like, 'No, no, no, that's not me, but I need to figure out what this is,'" he told me. But when he visited a friend in California for a weekend at a cabin, he had a meltdown. "I didn't know what a meltdown was then, and I just remember crying in his arms." By then he could see signs everywhere; he knew he had autism.

But Marble had trouble getting diagnosed because many doctors still do not diagnose adults and some will not accept insurance, which meant Marble had to save up for it. Finally, he spoke to a friend who also worked in the Obama administration that he was beginning the process but still needed some accommodations at work. Marble's friend suggested he meet with Ari Ne'eman, then an appointee to the National Council on Disability. When the two met for coffee at a Starbucks near Ne'eman's office in the Foggy Bottom area of Washington, Ne'eman noticed Marble was having trouble and suggested that they talk in his office instead, which was quieter. Marble realizes now that this was a setup by Ne'eman so that Marble could notice how his sensory processing differed in loud and quiet settings and help him accept he was in fact autistic.

As Marble waited to receive his official diagnosis, Ne'eman suggested Marble visit a meeting of autistic adults at a coffee shop that happened after-hours. It was there that he met autistic people like himself: lawyers, people in the military, and other professionals. Shortly thereafter, Marble wound up going to an all-autistic retreat in a cabin in the woods, where he met autistic people who could not speak.

"I remember crying the first time I was speaking with a nonverbal autistic who was communicating slowly through an iPad," he told me. "And I realized we experience the same world in much the same way, and yet the world is looking at us radically differently and I was looking at us radically differently."

Marble eventually left the Obama administration to work on Hillary Clinton's campaign, after which he moved to the Bay Area, but he had trouble finding work. A friend suggested he meet with a tech executive who was advising a program called Autism Advantage, which was run by an organization called Expandability (now called Neurodiversity Pathways) and focused on getting autistic people into integrated employment (a fancy way of saying autistic people working with non-autistic people).

Marble finally found his professional stride at Neurodiversity Pathways. He also worked with its Workplace Readiness Program, a six-week session that creates individual service plans, which, like individualized education plans (IEPs) for disabled students, make an "actionable plan" to help neurodivergent people across the spectrum find and set personal and professional goals. The program also helps them prepare for the interview and application process to ultimately find work. Afterward, they are offered postemployment job coaching and support. The program also conducts simulated work environments and Marble frequently took students around to companies so they could hear from professionals with the same skills they had.

Marble and I have many things in common: We're both former White House interns and autistic men, and we are both outliers. We have been fortunate enough to have had positive mentors and supportive employers. None of these things are guaranteed for the community at large. There are two dominant myths surrounding autistic people in the workforce. First, autistic people are often the victims of what former George W. Bush speechwriter Michael Gerson called the "soft bigotry of low expectations," meaning they are expected to be unable to work or only able to work jobs that pay subminimum wage. The second myth is the inverse of the first: people view autistic people as being hypercompetent in the fields of science, technology, engineering, and mathematics, as if we should all be coders in Silicon Valley.

In 2012, as professionals were debating whether to change the diagnostic criteria in the *DSM-5* for autism disorders, NBC's medical editor Nancy Snyderman, who was a physician, said that while some people believed that kids who had been diagnosed with Asperger's syndrome could be mainstreamed and didn't need as much help, others said that those were "exactly the kind of kids you should invest in because those children are the Silicon Valley of tomorrow."

But this binary abandons and actively harms many autistic people who live between these poles. Autistic people excel in a myriad of jobs. No one should presume to know what autistic people can do; what should be presumed is that autistic people belong in whichever professions they choose. That being said, autistic people's value and worth should not be tied to whether they are employable. It doesn't matter if an autistic person holds a high-paying job or receives government assistance; autistic people should be viewed with the same dignity that all people deserve.

My Career

In my family, politics were always discussed. However, I found my way into a career in political reporting thanks to my first love, music. My dad was a drummer growing up and his favorite bands were groups like Rush, the Cars, Led Zeppelin, Ozzy Osbourne, Van Halen (my dad and I disagree on whether David Lee Roth or Sammy Hagar was the better singer), the Scorpions, and AC/DC. What's more, my stepdad played bass and went to Woodstock but left early because he thought it was a dump. When my mom started dating him, he introduced me to Bob Dylan, Bruce Springsteen, the Animals, Jeff Beck, and the Rascals. My dad's and stepdad's love of music made me want to pick up a guitar, and I got one for my tenth birthday on the condition I take lessons.

My first teacher was an austere Russian who berated me for not practicing or playing right, so my mom hooked me up with a new one. JohnPaul Trotter was eighteen years old at the time and I remember him wearing khaki shorts and band T-shirts. He introduced me to the musicians whose names were on those shirts, people like Jimi Hendrix, Santana, Eric Clapton, and Stevie Ray Vaughan, as well as their influences, like Buddy Guy, Albert King, and John Lee Hooker. He even took me to my first concert: the Who at the Hollywood Bowl. That night, we somehow got backstage and wound up chilling with Slash, the long-haired and often shirtless and top-hatted (though not that time) guitarist from Guns N' Roses.

It was JohnPaul who noticed that I liked the stories behind music almost as much as I liked the music itself, so he suggested I get into journalism. I loved learning that Stevie Wonder wrote my favorite Jeff Beck song, "'Cause We've Ended as Lovers," as an olive branch to Beck after Stevie kept "Superstition" for himself even though it was Jeff Beck's futzing around on the drums that inspired the song. I liked reading which postwar blues guitar players influenced Eric Clapton (who loved Albert King and Robert Johnson) and Jimmy Page of Led Zeppelin (who liked Willie Dixon). Thanks to JohnPaul, I applied to work for my high-school newspaper. They needed someone to write up small pieces about the 2008 presidential race, and I caught the political bug then and there. Covering the primary race between Hillary Clinton and Barack Obama seemed like the most exhilarating job there could be.

Hence, when I came to Washington three years later to intern at the White House, the place was a natural fit. I loved that it was a city that thrived on politics and that I was praised for the fact that I lived and breathed my work. Being a White House intern mostly involves menial tasks, and I didn't do much of them in the White House even, let alone in the Eisenhower Executive Office Building. I worked at a remote building, sending out standard responses to letters addressed to Michelle Obama, her daughters, and their dog Bo. (I had a good laugh on my first day when I saw a stack of letters to be sent out to Eagle Scouts, not unlike the one I received from the White House when I achieved the rank.)

But despite the menial tasks like uploading letters into the computer system and determining what letter elicited what kind of response, I loved the gig; I loved the weekly meetings in the East Wing. But more than that, I liked the fact that I was working in a city and in a profession where my encyclopedic knowledge of politics was celebrated. This was readily on display when high-ranking aides to the president would speak to our intern class. I would research their past and try to ask probing questions. If they'd worked for Wall Street between the Clinton and Obama administrations, I asked if it affected their views on financial policy. I loved watching them be visibly uncomfortable. I liked asking Obama's speechwriter Jon Favreau about his writing style, and at our final Q&A with Michelle Obama, I got the first question. It was about why she'd left big law for public service, which led to a standing ovation. My buddy Eric Jones (now an Emmy-winning producer at *Good Morning America*) used to say with his sweet Vidalia onion accent from Georgia that I was always "rea-tee" (not just "ready") for these Q&A sessions.

But I occasionally disagreed with Obama's policy in Afghanistan and found myself trying to determine the difference between his foreign intervention in Libya and the War in Iraq, which was one of the reasons I didn't trust George W. Bush and the Republican Party. Those little breaks made me conclude that a career in Democratic politics was probably not for me. But I also knew I wasn't a Republican like many people in my high school. All of this showed me that journalism was a more natural fit.

That turned out to be the right decision. Between my junior and senior years at UNC, I interned at the *American Prospect.* Through that job, I befriended Jonathan Chait, an alum of the magazine, who introduced me to other political journalists like Jonathan Cohn, then an editor at the *New Republic.* Cohn liked my work, and while he couldn't offer me a job, he let me freelance for the magazine that I had loved for years. I parlayed that into other freelancing gigs before working at *MarketWatch*, *National Journal*, *Roll Call*, the *Hill*, and the *Washington Post*.

This is the point in the story where people expect a happy ending, the part where an autistic person goes from being a social outcast to channeling their special abilities into gainful employment. It's probably why a few people bought this book. This happy ending is seen in Temple Grandin when she pioneers new methods to handle livestock because she can "think in pictures." It's John Elder Robison in his book *Look Me in the Eye* learning to set off bombs in rock-and-roll band Kiss's guitars and then working on toys before moving to cars. It's every autistic person in Silicon Valley who can program a computer, no sweat.

These stories are neither bad nor untrue. In fact, they offered a much-needed corrective to the idea that autistic people could not work or that their autism meant they were defective. But highlighting hyper-savants who improve companies' bottom lines implies that only certain autistic people are worthy and deserve to find employment, and it positions autistic people who can't work without government services as a burden. Finally, it ignores the many autistic people who exist somewhere between those two extremes who have simply not been afforded adequate resources to find and keep a job.

But perhaps most important, these narratives put the onus on autistic people to find a super-skill that will make them an asset to employers rather than forcing employers to become more accepting of autistic workers. I was incredibly fortunate that I found something I loved as much as music when I became a journalist. I was even more fortunate that I had a family that fostered that passion and that I had the resources to pursue it. But not everyone has those luxuries; there needs to be institutional support for autistic people regardless of their work environments. Autistic people should feel their only limit to working in a profession is their own capacity to do the job rather than any outlying barrier that would exclude them.

Getting the Job

Because there is little accurate data about the number of autistic adults, it is difficult to track specific employment statistics about autistic people. One commonly cited statistic is that 75 to 85 percent of college-educated autistic people are unemployed or underemployed. Drexel University's National Autism Indicators Report found that paid, community-based employment, which is to say, autistic employees working alongside nonautistic employees, was, at only 14 percent, "the least common outcome" for autistic adults who received state developmental-disability services. And it isn't due to a lack of desire; a quarter of autistic adults had community employment as a goal.

This employment gap is partially what drives John Marble in his work. When we first met in 2018, he told me that when he speaks to companies, he emphasizes that hiring autistic people is not an act of charity. "There's a lot of companies that, for whatever reason, if you say autism, it's this sparkly thing that grabs their attention, and they'll open the door and say, 'Hey, I want to learn more, what can we do?,' not really knowing that they even need to employ autistic people," Marble told me. This is why Neurodiversity Pathways organizes introductory luncheons with companies to discuss neurodiversity and why autistic people are beneficial employees. "And that's not to say that we sell [autistic people] as some sort of superhero. It's just a different way of experiencing the world that's beneficial to a team," Marble said. He emphasizes that it is important to have many diverse types of people and notes autistic people offer certain skills. Marble said that while many neurotypical people have similar traits, autistic people often bring unique passion for projects that interest them, an attention to detail, and an integrity that can improve any team.

However, without proper support, there are many impediments that prevent autistic people from finding a job. For one, traditional job interviews often require unspoken rules and trick questions. Along with the daunting interview process, employers may have misconceptions about autistic employee's needs. For this reason, many autistic people who can pass as nonautistic often find it easier to not disclose. Marcelle Ciampi, the lead recruiter for Ultra Testing's neurodiverse hiring initiative, said she always regretted disclosing she was autistic to her employers.

"They don't know a lot of the strengths and gifts. They don't know that it doesn't cost that much money to have workplace accommodation or workplace adjustments," Ciampi, then fifty, told me in a phone interview. Ciampi, who used the pen name Samantha Craft for her book *Everyday Aspergers,* also said that once someone discloses, employers begin to question whether an autistic person is the best person for the job, whether this person can become a manager, and what challenges for the office this person would pose.

However, in recent years, multiple companies have made a push to specifically hire autistic people. A variety of companies, ranging from financial institutions like JPMorgan Chase and UBS to technology companies like SAP, Google, Salesforce, and Microsoft, have neurodiversity hiring initiatives. Many of them even include the same language the autistic community prefers and uses. For example, a statement on Goldman Sachs' website from April 2019 read: "Our goal is that this initiative encourages our people to share their stories and learn about how they can support neurodiverse individuals as managers, mentors, and colleagues inside and outside the firm." Many of the companies that have autism hiring initiatives recognize the challenges with job interviews.

The "Autism @ Work Playbook," compiled by companies like Microsoft, SAP, EY, and JPMorgan Chase as well as the University of Washington's Information School, says that interviews can be stressful and an undue burden on autistic job candidates but also that "traditional interview techniques may not always uncover the relevant skills the interviewee has." Some companies, like SAP, offer an alternative monthlong interview process that has candidates build robots out of Legos based on a set of instructions. Jose Velasco, an executive at SAP who has two autistic children, told *Inc.* back in 2015 that the goal of these exercises was to gauge how candidates read and follow instructions. Neurodivergent candidates are also allowed to pick the formats of their interviews; they can choose walking and talking or meeting one-on-one or in group settings.

What's more, these hiring initiatives exist in companies all over the world. Thorkil Sonne founded the Danish company Specialisterne in 2004 after he learned his son Lars was autistic, and he has pledged to create one million jobs for autistic people. When I spoke to Tara Cunningham, who was then the CEO of Specialisterne's U.S. organization, on the phone, she immediately mentioned how there were still plenty of misconceptions about autistic people in the workplace, including the idea that many autistic employees love repetitive tasks, which she dubbed "bullshit."

Cunningham said there was an assumption that autistic people would be in the exact same role for the rest of their lives. She also said she was concerned about new trends, among them that as companies see the benefit of autistic employees, they will start poaching them from other companies.

"Right now, we can say, 'There's awesome retention rates,' but pretty soon, autistic people are gonna get headhunted like everyone else and possibly harder," Cunningham said. Such a trend could have unintended consequences, including undermining the idea that autistic people are loyal employees.

The Loyalty Trap and Other Issues with Corporate Hiring Initiatives

The concept of loyalty as an autistic trait permeates many hiring programs. One story by my former employer *MarketWatch* said that "being honest and loyal" made autistic employees an asset—though I clearly failed on that front when I quit working for them after four months. But while the idea that autistic employees are loyal may be based in reality, it is by no means universally true. Steve Lieberman, who is autistic and used to work for Senator Robert Menendez, put it best when he tweeted that disabled employees were not loyal; rather, "We're stuck."

"I worry about saying, 'Hire an autistic person. They're yours for life.' I don't believe that's true," Tara Cunningham said.

Another one of her concerns is that autistic people are considered good assets only because they are perceived as more productive. In countless news reports, increased productivity is one of the features employers highlighted when talking about hiring autistic people. But Cunningham worries that this could backfire.

"When people say that their autistic employees are a hundred and seventy percent more productive than their neurotypical counterparts, now that's a huge problem," she said. If autistic people are only slightly more productive, employers feel cheated for getting someone who is only, say, a hundred and forty percent more productive. The productivity myth can also make neurotypical employees resentful of autistic employees.

There is also concern that these types of programs are focused on hiring autistic people into STEM fields only. Julia Bascom, the executive director of the Autistic Self Advocacy Network, said while certain companies do a good job and promote autistic people in management, there is a risk of just focusing on a certain type of autistic employee.

"[These programs are] very focused on tech, which causes a lot of problems, because, while statistically, we may be represented in certain [STEM] majors, but a lot of autistic people have learning disabilities that affect math or other things like that," she said.

John Marble agrees with this sentiment, explaining that as of right now, autism hiring is concentrated in one "slice," as he calls it. At the same time, parents of autistic kids know their children have the ability to work in any field, from communications to groundskeeping to food service to graphic design and so on. Marble said the narrow focus of hiring initiatives becomes frustrating.

And I understand this frustration, as I'm a perfect example of an autistic person who would be overlooked by certain companies' hiring initiatives. I am utterly terrible at math—when I took a computer coding class, I got a C-minus (this despite the fact that I flunked most, if not all, of my exams but always finished my homework thanks to the tutoring center on campus). Suffice it to say, I never had a desire to work in science or technology. And many other autistic people have no desire to do so. There are some companies, like Ford Motor Company, that have already taken these initiatives beyond the normal STEM sectors. But hiring practices that are suitable for autistic people should be adaptable for employers of any sectors. Chances are, there are already autistic employees in those businesses. It's just a matter of making them and future autistic employees feel welcome.

Bascom also said she was concerned that many of these companies were just siloing autistic employees and creating de facto sheltered workshops wherein autistic people were separated from the rest of their coworkers. Sometimes, they were even paid less. Bascom added that many companies used autism-at-work programs to hire autistic people with lower support needs as a way to avoid hiring autistic people with intellectual disabilities who might need more accommodations. It is why, after initially discussing a partnership with Specialisterne in March of 2018, Bascom said, ASAN withdrew from the discussion—it was clear there was no commitment to helping autistic people who had higher support needs.

Alan Kriss, who is the head of Specialisterne's North America branch and served as head of the organization in Canada, confirmed that talks broke off at the time but said that the company has progressed.

"From our perspective, our objective and our focus is on helping employers to access talent pools of people who are in a position to or can be in a position to achieve competitive employment," Kriss told me in 2020, adding that Specialisterne doesn't focus on whether a person has intellectual disabilities and that almost all of the people recruited for work have some type of job coaching.

"We take people's strengths, and we leverage their strengths," he told me. "And as long as the challenges that they face aren't ones that are going to stop them from achieving competitive employment, then we're going to support them into that." Kriss added that the organization is trying to change "how the employment market operates."

"We're focused on helping employers to see that there are better ways of assessing people's talents and competencies," than just glad-handing and job interviews, which Kriss said has been shown to be a poor method for recruiting for neurotypicals too. "And just good management is going to get a lot of people success, [it's] going to help a lot of people be more successful," regardless of their neurotype. He also added that the company is moving beyond technology and financial services and hopes to work with the "intellectually challenged" but that there will be some hurdles because those aren't the markets where Specialisterne started.

Dignity

Even if there is success with hiring programs, the push to create autism-at-work programs can create tiers of worthy autistic people. This concept, of course, is not new. It goes back to the idea that there are high-functioning and low-functioning autistic people, which runs the risk of creating a power imbalance. It sets up the false narrative that autistic people who cannot find full-time work or who do not have the capacity to work full-time are not worth the same amount of attention or resources that autistic people who can become assets for companies receive.

Because of this, few autistic self-advocates think all autistic people will achieve full, integrated employment, acknowledging that there will always be a need for a robust safety-net program.

"While it's good to advocate for decreased barriers to employment for autistic people who want to work, there will always be some of us who simply cannot hold jobs," said endever star, a nonspeaking autistic person I met at Autspace, a retreat for autistic people.

"We need to consciously avoid saying things like 'autistic people can work too if we just have the right supports!'" star added. "That's a sweeping generalization that denies the reality of part of our community."

Star points out that even America's social safety nets create a dichotomy between earned and unearned disability benefits in the difference between Supplemental Security Income (SSI) versus Social Security Disability Income (SSDI). Disabled people with limited income often receive SSI, which is paid for by "general funds," like personal income tax and corporate taxes, whereas workers' contributions to the Social Security trust fund pays for SSDI and is based on their earnings. A total of 383,941 autistic people received SSI in 2019. These different funding streams reflect how America constructs a contrast between "deserving" and "undeserving" poor. American culture perceives recipients of SSDI as "earning" their income because they paid into Social Security.

"Meanwhile, people who have never been able to work or haven't worked 'enough' are given ONLY SSI, which leaves them in inescapable poverty for potentially the rest of their lives," endever star said over e-mail. "This is a very blatant expression of the way society views access to supports—there's an idea that we have to earn our supports or prove that we're worthwhile human beings in order to access them."

To add another layer of difficulty, the process for obtaining SSI benefits is baffling and as discouraging as possible. Andrew Savicki, the political prognosticator from Apex, North Carolina, said he hasn't been able to acquire SSI benefits. "One of the times they told me I wasn't disabled enough to get benefits because it is physically possible for me to hold down a full-time job," he said. However, not having a college degree and not being able to drive makes it extraordinarily difficult for him to find work. "So that's really hurt my independence [not] being able to get out there and find a job," he said in reference to not driving.

In my opinion, Savicki has one of the better political minds I know, better than a lot of the gasbags who wind up talking about politics and congressional races on television (including the gasbag who wrote this book). But the barriers to entry are still too high for him because of his inability to drive and his lack of a college degree. Without SSI income, it is difficult for him to have full independence, yet because of society's expectations of autism and neurotypicality, he is stuck in an income purgatory even though he wants to work. Savicki's predicament shows how the concepts of high-functioning and low-functioning hurt autistic people. When people exist somewhere between these two ideas, they often wind up getting neither accommodating work environments nor income from social safety-net programs.

Unfortunately, the default for many autistic people is not much better, since they can wind up placed in sheltered workshops, which are environments where disabled employees are separated from the rest of the population except for a nondisabled supervisor. Furthermore, they are often paid below minimum wage. The Fair Labor Standards Act of 1938 established the cornerstones of American labor, like the federal minimum wage and overtime pay. President Franklin D. Roosevelt considered it some of the most important legislation he signed next to the Social Security Act. Section 14(c) of the law allowed employers to pay disabled people "whose earning or productive capacity is impaired by a physical or mental disability, including those relating to age or injury, for the work to be performed" below the federal minimum wage.

This was the case with Maxfield Sparrow. When Sparrow—who is transmasculine and for the purposes of this book will go by *they* and *them*—was living in Kentucky, they had trouble staying employed, so they went to a temp agency, then a government agency, and then to a facility that processed pieces of foam that were used for floor mats. When the pieces came out of the mold, they looked like puzzle pieces, and it was Sparrow's job to take off the excess parts.

"This is a horrible job, and we couldn't even sit down," they told me. "We had to stand there and take these pieces off these things, and I wasn't allowed to use the bathroom except during designated breaks."

The place was sweltering hot, and the workers were not allowed to speak to one another. Ultimately, Sparrow couldn't continue working there because of the conditions. "I couldn't pay rent on what I was making, and so I couldn't keep a place to live," they said, noting that they were sleeping in a park because of this.

Thankfully, there has been a push from both sides of the political spectrum to get rid of subminimum-wage labor. For instance, Texas governor Greg Abbott, a Republican who uses a wheelchair after an accident, signed legislation championed by Republican legislators that required nonprofit organizations that employ disabled workers and had contracts with the state to pay them at least the minimum wage.

Still, sheltered workshops have many powerful defenders. During labor secretary Alex Acosta's confirmation hearing in 2017, Senator Maggie Hassan, a Democrat from New Hampshire whose son Ben had cerebral palsy, asked Acosta what he thought of subminimum-wage labor for people with disabilities. Acosta said that while he supported the authority of states to end the practice, the issue was "difficult." He said, "The very phrase 'subminimum wage' is a disrespectful phrase," rather than calling the idea of disabled people not receiving a fair wage disrespectful. And similarly, the National Council on Severe Autism [NCSA] has said closing these workshops shuts off an opportunity for disabled people.

"The idea that everyone with autism can achieve competitive, minimum-wage employment given the proper training and supports is pervasive in the disability community," the NCSA's webpage said. The debate is another example of parent advocates' interest (NCSA says it speaks for "the disabled who have no voice") opposing the interests of autistic self-advocates. NCSA calls these sheltered workshops "vocational options," although many times they are not the options of disabled or autistic people but the desired choice of parents.

Other groups, like VOR (formerly known as "Voice of the Retarded"), also vocally oppose eliminating subminimum-wage labor. The group calls the wages "specialized wages" that are "appropriate to their level of productivity" even as it admits the wages are lower than the federal minimum wage. Julia Bascom notes that these groups are often better at crafting compelling narratives for lawmakers, particularly during the Trump administration, which, compared with the Obama and even the George W. Bush administrations, lacked experience working with disability advocates.

"Despite being organized and loud, they're also the minority opinion," said Julie Christensen, the director of policy and advocacy for the Association for People Supporting Employment First (APSE), a group that supports disabled people finding full-time integrated employment. Conversely, the paradox of supporting integrated employment at or above the minimum wage is they are out living their lives and working, which makes it harder to have them coalesce and advocate for integrated employment.

"We don't necessarily have ongoing contact with them. That's the way you want it. They're not reliant on the system," Christensen said. This means it's harder to organize people who have transferred to competitive and integrated employment in the same way as parents who support these workshops. Christensen says this does not mean their concerns aren't legitimate, "but I think sometimes the volume makes it appear that it's a bigger political force than it actually is."

NCSA points to the fact that in states like Maine that have shuttered their programs, as many as two-thirds of people who worked in them remain unemployed. But the same case study NCSA cited—which was conducted through the George Washington University's Milken School of Public Health's Department of Health and Policy Management—notes that many people with intellectual and developmental disabilities who left sheltered workshops went into "community-based non-work services," such as skills-building and socialization services, field excursions where they interacted with nondisabled people. The same study also noted that people working in integrated settings declined as well.

In the same respect, places like Vermont, which ended funding for new entrances for sheltered workshops and made heavy investments in supported employment, saw 80 percent of people who worked in sheltered workshops transition to supported employment, while the rest went into a nonemployment community-support systems. Similarly, Vermont has a 38 percent integrated-employment rate for people with intellectual and developmental disabilities, which is double the national average of 19 percent.

The NCSA website said that "wage-earning is not the primary purpose of these places," implying that the there are other valuable benefits to participants, such as "Medicaid-funded supports, in-home assistance, residential care, behavioral support, respite, recreation, and other therapeutic services."

A fear of a loss of Medicaid services is a legitimate concern. But as Ari Ne'eman, the founder of ASAN, wrote, in 1980, Congress passed the Social Security Disability Amendments Act, which included section 1619(b), which was made permanent in 1986. The program allowed recipients of SSI to continue to access Medicaid even when their income rose above the cutoff level.

Bascom says one obstacle to ending subminimum-wage labor is that many times, sheltered workshops let caregivers get out of the house and provide transportation to and from work, which is not always the case with customized or supported employment. "It could be, but that sort of ease isn't always there." One reason for this may be that subminimum-wage labor has been the default for more than seventy years. Bascom says adding these services to integrated and competitive employment needs to be made easier.

At the core, though, the arguments for subminimum wage still devalue the work of disabled people. While it is important to offer respite to caretakers and all these supports are valuable, none of these arguments deal with the fact that disabled and autistic people's work has worth and is no less deserving of adequate compensation. There can and should be means of respite care and day programs that are fully funded and that are integrated into the community. But these arguments still ignore that autistic people—and disabled people as a whole—have as much worth as other people.

Public policy seems to be moving more toward supporting integrated employment. In 2016, both the Democratic and Republican Party platforms called for ending subminimum-wage labor (though they persisted through the Trump administration). In September 2020, the U.S. Commission on Civil Rights said that Congress should pass legislation to repeal the 14(c) with a phase-out period to enable transition to work. The commission also recommended that Congress spend more for supported-employment programs.

"Given the volume of concern and the experience that we've seen among some states with exiting the program, we thought it was important to make sure that there is a phase out and that appropriate supports are in place," Catherine Lhamon, the chairwoman of the USCCR, said. Lhamon said that her own view was that the most significant element to success was to believe that disabled people could succeed, but "we now operate a program that gives permission structure to not believe in the possibility for success."

A big part of that was due to the fact that many of these structures were built when people didn't properly understand autism or intellectual and developmental disabilities. The year Roosevelt signed the FLSA was the same year that Leo Kanner began his survey of eleven autistic children. And, of course, it will take even longer to change the paradigms we have now. The new knowledge over the years coincided with changed expectations. Neil Romano, who was appointed chairman of the National Council on Disability by President Donald Trump, testified during the USCCR's deliberation that in the 1930s, paying disabled people pennies on the dollar was considered charity "because a lack of belief in people with disabilities was the game at that time."

But autistic people, and all disabled people, do not need charity, which comes from a point of pity (though there is a place for it). Rather, they need justice, which is born out of the American ideal that people have inherent value. This is something the generations that came up around and after the passage of the ADA and IDEA take as a given, Christensen says.

"We have a generation of people with disabilities who have grown up with more options and they are not going to self-select into a segregated environment when their entire lives, their K-through-twelve schooling, has been based on an inclusion model," she told me. "And so, while people won't say it, there is writing on the wall and a recognition that these types of programs are going to die by attrition eventually anyway."

The divide between parents and self-advocates about sheltered workshops and subminimum-wage labor is about what is most valuable. Parents tend to value the accoutrements of subminimum-wage labor—the respite care, the chance for their kids to get out of the house, and the accompanying Medicaid services—and they don't want to lose it. But autistic people and self-advocates want to emphasize their labor is equal in value to nondisabled labor and so it deserves the same amount of compensation. Autistic people being paid below minimum wage sends a message that they can cosplay as workers and have small concessions, but their labor will always be considered subpar. The illusion of fairness will never be a substitute for the real thing.

Accommodations in the Office

Although on paper, I have definitely had a lucrative career, it has been laced with a sense of furtiveness. In college, I didn't initially disclose I was autistic, and when I got my first job at *MarketWatch,* I decided not to disclose.

I was making good money, but I was covering financial transparency, a beat in which I had little to no experience. That meant that when news broke, I was overwhelmed, trying to simultaneously learn the jargon of the beat while filing my stories on time. I would frequently make errors because I was so focused on trying to get the little details right that I would make careless and easy mistakes, like not giving someone's first name the first time I mentioned them.

I was frequently embarrassed, and I felt like I had let down the people who had hired me. Thankfully, my bosses Steve Goldstein and Jeremy Olshan, among others, were great. They took a risk hiring a kid who had no business doing that job, and they gave me chance after chance after chance. It was my fear of getting found out, my difficulty adapting to a high-pressure news organization, and my lack of knowledge that ultimately led me to burn out.

At one point, I was so afraid that they would fire me, I thought about killing myself before they could because I didn't want to disgrace all of the people who had put their faith in me. This, combined with spinning out in my personal relationships, was a catalyst for me to begin therapy. (I will talk more about my mental health as an autistic person in subsequent chapters.)

Still, the pressure was so great that when *National Journal* offered me a position, I took it, even though I had been at *MarketWatch* for only four months. I felt ashamed to leave and felt like I'd failed, but I knew that job was unsustainable for me. Through no fault of my employer's, I had burned out. Just like at Chaffey and UNC, it was only when I acknowledged my need for accommodations that I began to really succeed.

It was at *National Journal* that I became comfortable disclosing that I was autistic, and I even wrote the piece that became the launching point for this book about being an autistic reporter in Washington. I thrived at that job where I got to cover my first love, politics. Even though I had gotten a C in economics in college, I found myself enjoying writing about economic policy. I also enjoyed going to Capitol Hill and interviewing senators, some of whom wound up running for president. One of the moments that I knew I was where I was supposed to be was in 2015, when I was covering an event with Vice President Joe Biden and Senator Elizabeth Warren of Massachusetts at the Washington Hilton, the same hotel where my parents, my sister, and I had stayed when we visited Washington when I was ten.

I eventually left *National Journal* to work at *Roll Call* and then the *Hill,* mostly on breaking news, which, once again, brought its own pressures and fears of working in high-intensity jobs. The difference was, I could no longer hide the fact that I was disabled. This meant I could have a work environment that allowed me to succeed. It removed the tension and fear of being "found out" that many other autistic people have. Even at my job at the *Washington Post,* my then boss Adam Kushner asked at our first lunch if there was a way to accommodate me. Unfortunately, bosses like Adam are a luxury that many autistic people don't have.

The lack of a pipeline to media jobs for autistic people means many suffer without the opportunities I have been given. Every few months, I get e-mails from autistic journalism students or interns at news outlets who ask for advice about how to break into the industry. I am more than happy to help them, but there is still a huge lack of neurodiversity in the communications field. For whatever reason, people don't see media as a field autistic people would want to enter. A potential explanation is that journalism—with its constant demand of interacting with people socially through interviews and edits—is not seen as a typical autistic career.

But the desire and the demand exist, just as it likely exists in every sector. Autistic people work not just in technology or finance or even journalism but in every industry and at every type of income level. But because neurotypical people tend to think that autistic people are either unable to work or must be savants who understand computers better than they understand people, plenty of autistic people get ignored. The myopic and limited stereotypes often wind up hurting those autistic people who exist in every sector and lock out autistic people who want to enter various parts of the economy.

Retention and Inclusion

But even if autistic people are hired through these autism-at-work programs, they still desire the same thing any employee wants: inclusion. Ciampi, the recruiter mentioned earlier in this chapter, said that there is a big difference between diversity and inclusion and that many companies with autism-at-work programs are focused only on getting autistic people hired.

"Inclusion measures after they're through the door are really essential. As well as before they're through the door," she said. Part of that includes tailoring the entire hiring and screening process with the input of neurodivergent people, modifying the interview process not just for autistic people but everyone, and including autistic people in mainstream types of hiring so they're not being singled out from other employees.

"Let's look at how we can put support systems in place for everyone and not make it about 'How can we include autistics?'" Ciampi said. "Because as soon as we start saying, 'How can we include autistics?,' we're singling them out [as being] in need of help, more than another minority or more than another human being."

John Marble said the dichotomy for many employees at companies with autism-at-work programs is that they are either "crashing and burning" or "growing and leaving."

"So that's the pattern that I've seen is that these companies with their great incentives have gotten really good at hiring for their programs for their departments," he said. "The established companies are able to fill their order of autistic people and they can help train them, onboard them. What I don't see is a real deep understanding of autism from the autistic point of view."

This isn't to say that autistic people don't have real needs. But it is important to recognize that everyone has some sort of need. "If we can address all those issues, and then put in place things that help the whole workforce, such as handbooks, and procedures and policies that reflect inclusion for everyone," Ciampi said. This could be as simple as having monthly meetings with no agenda except to talk about improving the workplace, noting that one of the primary issues all employees value is having a sense of belonging in their job.

"That is inclusion. So, let's look at, how do we build a sense of belonging? And start from a very broad focus, and then go down narrow," Ciampi said.

One company that is working on becoming more inclusive of autistic employees is Square, the financial services and mobile-payment company. That began when Chris "C. J." Ereneta, who has an autistic son, wore a shirt supporting neurodiversity to work. The shirt attracted the attention of Chris Williams, who is autistic and married to an autistic woman with whom he has autistic children (more on the Williams family later). Square has several employee resource groups (ERGs) for different identities, so Ereneta, Williams, and a third colleague, Mary Overbee, came up with the idea to create one for neurodiversity.

Initially, few people participated in the group's Slack Channel. But Ereneta said the group changed its focus to include the entire gamut of neurodiversity, not just autism, which has helped it grow.

"I think we're getting more people who are stepping out and claiming either an autistic identity or talking about their struggles with anxiety or depression or attention," he said. "We have a lot of activity around attention differences."

One of the group's members, Claudia Ng, who considers herself neurodivergent with ADHD traits, said the experience has been transformational for her since she long suspected she might have some attention issues but ignored it for most of her adulthood. But having Ereneta as a boss and identifying her differences and how they are beneficial led her to accept her differences.

"[Ereneta] made it clear to me that there are differences, and they're not necessarily good or bad, but they're just different, and that might influence the way that my teammates might work [and] would most likely be different than the way I work, and so for me it's self-actualization," Ng said. She added that some of her special skills include pattern recognition, which has been seen as a trait of some neurodivergent brains. Similarly, in the past, she would frequently lose interest in things, but the inverse was that she would find interest in a wide array of subjects all the time.

Generally, Williams said he found Square to be more welcoming and accepting of him than previous employers. In our first interview in 2019, Williams broke down his career into three distinct moments, starting with his time before working in finance, when he delivered pizzas, managed a university coffee shop, and waited tables. He fell into finance "by accident," at E-Trade, where he climbed the corporate ladder before working for Morgan Stanley for a year and a half. But while Williams was successful in getting jobs, he struggled with the structure and management style within the company. One example that sticks out for Williams was when he spoke to a manager about professional growth, and the manager told Williams that he didn't "eat the right way" with utensils. The manager said the company would not put Williams in front of clients for this reason.

Later, Williams's wife had a series of complications with pregnancies and could not return to work, so their income was suddenly slashed by half. This led to credit complications, and he and his family eventually lost their house.

Williams found a job in consulting and moved to California, where he met his future manager at Square. There, Williams decided to disclose his autism, which he said has been liberating.

"I really felt a veil strip away, or a layer of difficulty strip away here [at Square]," he said. "Other places have been much more difficult, where places haven't necessarily understood why I've communicated the way I have or why I've struggled the way I have."

Tyneisha Harris, the technical recruiter at the company, said that she was interested in finding better ways to recruit neurodivergent people—for instance, by making sure proper language is used in job listings and training managers to be mindful about accommodations. At the same time, Ereneta said he wants to make sure Square is a good place for employees who are already there.

"I'm one of the voices saying let's focus on the inclusion part, let's focus on continuing to make Square a great place to work for the people we already have here who are different," he told me.

"Yes, the recruiting pipeline, those practices, we have to adjust those," Ereneta said. "But with all sort of diversity inclusion issues, if you only solve for the diversity recruiting part and people arrive and it's not a company that supports you and your differences, that's where a lot of my attention is."

From Struggling Employee to Business Owner

When John Marble began working, he didn't have supports or accommodations as an autistic employee, partly because he didn't realize he was autistic and partly because they did not exist at the time. Autistic people were not seen as being able to work or thrive in workplace environments. That makes his ability to succeed in politics and government all the more remarkable, but it's also why he is trying to focus on supporting autistic employees.

A year after I first met with him, Marble and I caught up to discuss his new company, Pivot Diversity. As Marble continued to consult with Neurodiversity Pathways, he said he inevitably heard from people who were worried about their autistic coworkers struggling. Employers also asked about the hiring process or their benefit packages. But when they asked Marble who addressed these needs, there was no one comprehensive organization.

Once again, much like the spark that it took for him to accept his autism, it took a friend telling Marble that he was an employment and workplace expert to give him the push to start his own company.

"I realized that the current existing autism work framework that exists, which is very well intentioned, is a lot to get a company to buy into," Marble said. "It requires selling a company [on] what autism is, then the value of autistic workers—not to mention the value of diversity. Afterward, it requires learning how to adopt the program, sourcing and retraining people. All of this makes for a costly financial investment that discourages companies.

"Meanwhile, you've got autistic people in almost every company who are struggling, who are drowning, who are trying to make it somehow," he said. "I realized that if we help the people in existing companies, and if we help their companies become more knowledgeable about autism in the workplace, that is a much lower bar to cross."

He noticed that even though he worked with cohorts of talented autistic students while at Neurodiversity Pathways, they soon got weeded out for jobs in Silicon Valley because of screener calls. Marble wondered how to bypass this obstacle.

"That's when I realized that, oh, if I started a company to help companies support their existing employees, then that starts to grow a culture within that company that's going to be prepared for the young adults that I teach but also for a lot of other people," he said. When a friend asked him what his biggest pie-in-the-sky goal was, he said he wanted to put autistic people at the center of programs meant for them, which meant putting an end to well-meaning people who made decisions based on incorrect assumptions.

Ultimately, Marble said he was confident that if he started a company centered on autistic people, it would be able to provide better solutions than ones coming from nonautistic people. It was through that discussion that Pivot Diversity was born.

"Pivot is doing this work in a way that helps companies support their neurodiverse employees, which builds a sense of inclusion and belonging for all employees, which I think is very dynamic, because we just don't leave it to autistic employees," Marble told me. "We might be the hardest case, but if you do these practices with autistic people, other populations in your company are going to thrive."

The more I talk to and interview autistic people, the more I am reminded of an analogy that Marble often uses—he likens being autistic to being French.

"There's millions of different ways to be French, and a gay fashion designer in Paris and a Catholic nun in Bordeaux are going to be radically different," he said. "But they still understand each other as French."

Marble notes that if you put autistic people in a room without any neurotypical people, their communication radically alters. I have noticed this in my own interactions with autistic people. We feel less fear and anxiety about stimming; we allow each other to ramble about our favorite subjects. We indulge each other and go back and forth, sometimes in conversation and sometimes literally rocking back and forth.

In the workplace, autistic people generally do not have this luxury of understanding. Some employers' expectations are set too low; they don't think we will be worth the effort expended on accommodations. Just as problematic, some employers set the bar impossibly high, expecting autistic workers to be their secret sauce. Rarely are we afforded the chance to be normal, to navigate workplace politics, to be ourselves and simply contribute as we see fit. Like in so many different arenas, autistic people at work are constantly adjusting ourselves not to our own expectations but to the expectations of the outside world.

4

"Gimme Shelter"

HOUSING

Julia Bascom has advised presidential candidates and regularly advocates for autistic people's rights on Capitol Hill. Some may think that means that Bascom, who serves as the executive director of the Autistic Self Advocacy Network in Washington, DC, is more high-functioning than Leo Rosa, an autistic person with limited speaking capacity who lives in the San Francisco Bay Area of California, but their stories showcase just how reductive and futile functioning labels really are.

Rosa, who was nineteen when I interviewed his mother, Shannon Des Roches Rosa, continues to live at home and requires 24/7 support. Some parents of autistic people with higher support needs say that the gap between Bascom's and Rosa's autism is as wide as the country between the two coasts they inhabit. There is a commonly held belief among the general public that because Bascom does not live with her parents and is capable of speech, she is high-functioning, and because Leo requires around-the-clock care, he is low-functioning.

But that would ignore the fact that Bascom herself is able to do her advocacy work because of the help of her support person, Colton Callahan, who has lived with her for roughly seven years. In the same respect, the fact Leo needs his family's support most of the day does not diminish his right to pursue his happiness or mean that his life is any less valid than others'.

Bascom and Rosa prove how fickle the concept of functioning labels is and show how independence for autistic people can manifest itself in multiple ways. Unfortunately, just as "experts" in the twentieth century thought it best to keep autistic people in institutions, there are still people today who support this dated approach. This takes away autistic people's opportunity to live within their community and with their families—something we know improves their overall well-being.

The living situations of autistic people before the twentieth century are largely unknown because their condition had not yet been documented. As Jack Pitney notes in his book *The Politics of Autism,* the inaugural issue of the *Journal of Autism and Developmental Disorders,* which was then called the *Journal of Autism and Childhood Schizophrenia,* opened with the line "One is entitled to wonder: What happened until recently to unfortunate children who, through no fault of man, were condemned to sufferings now belatedly recognized as psychotic ailments?"

The autistic people in the twentieth century who were misdiagnosed were relegated to institutions and state hospitals. David Mandell, a professor of psychiatry and pediatrics at the University of Pennsylvania, found that 10 percent of 141 residents at Norristown State Hospital outside Philadelphia had undiagnosed autism and all but two had been diagnosed as having "chronic undifferentiated schizophrenia." Typically, when children received a diagnosis like this, their parents sent them to institutions, often even removing their pictures from the house. As Roy Richard Grinker noted in his book *Unstrange Minds,* Bruno Bettelheim said surveying kids for all hours of the day could teach him which treatments were most effective at "thawing out" children who had been frozen by their mother's "black milk."

After Bettelheim's death, writer Richard Pollak wrote that Bettelheim subjected his children to strikes and whips with a belt. One former resident at his institution, Alida Jatich, compared him to a cult leader who bullied everyone from his students and residents to the children's parents. Another, Ronald Angres, said Bettelheim "insulted people just in order to break any self-confidence they might have" and that he "thoroughly broke mine," as he also used corporal punishment against the children at his institution, despite his preaching against it in public.

Autistic people were not any safer at other facilities, and scores of them were subjected to truly horrid conditions.

Cal Montgomery, who was fifty-two when I interviewed him, is old enough to remember the realities of these institutions. Montgomery's father was a diplomat, which meant his family moved around frequently, and he spent only about one semester in two different American schools before he was written off as "emotionally disturbed." Despite this, he was able to graduate from the University of Texas. Still, as an autistic person living in a society that didn't yet fully understand his condition, Montgomery was institutionalized at Austin State Hospital in 1987.

"I did something really stupid in public and the police picked me up. They stood behind me with guns and said, 'We can do this the easy way or the hard way.' And I cheerfully signed myself in voluntarily," he said with more than a hint of sarcasm.

Sadly, according to Montgomery, that would be "the best institution I've been in." He was released after fifty-nine days. Soon after, he was institutionalized at McLean Hospital in Belmont, Massachusetts, for two years, with a series of short-term stays. Between 1989 and 1992, he was sent to Wild Acre Inns, first in Brookline, Massachusetts, and then in Cambridge. In 2017, Montgomery spent three short-term stays at Warren, Barr, and Kindred in Chicago.

"This second institution [in Massachusetts] literally, they tortured me," he told me in an interview in 2019. He was referring specifically to their use of electroconvulsive therapy (ECT). "It was framed as an option, but it was not an option," he added.

Montgomery said that finding appropriate housing for autistic people is much more difficult than it is for people with physical disabilities. "I personally don't even have the language yet to talk about what a socially accessible environment would be like. So, it's hard to advocate for it because I can't explain it to anybody," he said.

Many people on the spectrum were able to avoid the fate of being sent to institutions because they did not exist within the narrow confines of autism's definition back then. However, these same people still had difficulty thriving because they lacked the proper support.

Samantha Crane, Julia Bascom's colleague at the Autistic Self Advocacy Network, said one reason for this is the belief that autism is fundamentally different from other disabilities.

"Instead of learning from the experiences of other groups, a lot of parents and service providers are still trying to re-create models that other populations have already tried and found ineffective and overly restrictive," Crane said.

A lot of the public's misunderstandings about autism are thanks to fictionalized tropes. Historically, Hollywood has helped perpetuate narratives that infantilize autistic people. For example, the first time that many Americans were exposed to autism was through the 1988 movie *Rain Man,* in which Dustin Hoffman plays an autistic savant who goes on a road trip with his cold and calculating brother, played by Tom Cruise. At the end of the movie, Hoffman's character, Raymond Babbitt, returns to the institution where he started.

But as Steve Silberman's *NeuroTribes* pointed out, two of the autistic men who served as inspirations for Hoffman's character, Joseph Sullivan—the son of Ruth Christ Sullivan—and Peter Guthrie, were not institutionalized. It was likely that they were able to develop the very skills—such as Peter Guthrie's capacity to learn multiple languages and Joe Sullivan's near-photographic memory—that Hoffman portrayed because they lived around people who loved them.

In fact, the initial plan for the movie was to end it with Raymond Babbitt moving in with his brother, Charlie. This proved to be a step too far for autism "experts." Bernard Rimland, the cofounder of the Autism Society of America, advised on the movie, and he was adamant that state homes were the only appropriate setting for autistic people. Darold Treffert, who was the world's leading researcher on autism at the time, was also consulted for the movie. In his book *Islands of Genius,* he wrote that "the 'happy ending' in the original script is simply not realistic."

This advice came as a great disappointment to the film's original writer, Barry Morrow, who said, "I felt betrayed politically, but artistically, it was a triumph."

Silberman told me in an interview that he was overcome with a sense of irony when he spoke to the people who made the movie. "'Oh, why don't we hear more from autistic adults from the fifties?' They were in institutions, basically," Silberman said.

Even though Rimland did not institutionalize his autistic son Mark, he excoriated attempts to integrate autistic people into the community. In the editor's notebook page of *Autism Research Review International* from 1993, he used the term *advozealots* for people who supported "the handicapped" but were "in fact zealous advocates for their own Alice in Wonderland ideology"—aka people who, in his view, were too idealistic to act in the best interests of disabled people.

"Now they are destroying the institutions needed for the most severely retarded and autistic people," Rimland wrote. "Under the banner of 'empowerment' and 'human rights,' and 'full inclusion,' they have also set out to destroy the special education system created by decades of advocacy and hard work on the part of the families of mentally handicapped children."

At the heart of these debates is the question of who knows what is best for autistic people. For years, clinicians were the ones who urged parents to send their children to state hospitals and institutions, insisting this was in their best interests. But many of these institutions were rife with abuse or negligence. Parents' subsequent push to remove their children was an effort to take back and reclaim their authority after years of being told they were the problem.

These days, there is a divide between some parents and self-advocates over what the best environment for autistic people is. Many parents of autistic people with higher support needs have argued that their children need environments that are more restrictive—places that some self-advocates would deem institutions but that these parents believe can ensure their children's well-being and security.

Conversely, many autistic self-advocates see being a part of a community—whether it is living with friends, parents, a home-care worker, a roommate, or simply by oneself—as being part of the social fabric. It means our fate and our health are tied to others and we can't be relegated to seclusion.

Life Beyond Institutions

When we spoke over video, Leo Rosa seemed to be a generally happy person, and he greeted me when his mom asked him to say hello. He has a good life with a strong support system, which is crucial, since, as Shannon told me, he can't do anything independently.

"He has to have somebody with him in the house, because he doesn't know how to prepare his own meals, although he can do part of it," she said. On top of this, Leo can't drive, and if he went into the street, he would not necessarily know to stay away from passing cars. Leo can also be very dysregulated, which means he has difficulty modulating emotional responses. Shannon said that right before I called, he was having a meltdown because his schedule had been disrupted.

"He had a hard time waiting [for lunch], and I could see that if somebody didn't understand autism and autistic regulation and how easily a person can become dysregulated, that they might think that this was an awful thing," she said. "But I'm like, 'Well, you give him his lunch and let him process, and then he'll be al' right.'"

While Shannon is explicit about Leo's difficulties, she doesn't make herself out to be a victim (an all-too-common refrain in parents of autistic children). Rather, she is empathetic and knows none of this makes Leo any less human. He may have higher support needs than some other people, but they do not diminish his worthiness. Leo deserves all the respect and dignity Shannon's other children have, and understanding these needs allows the Rosas to know how to best assist Leo and themselves. As I heard Shannon speak, I was struck by the similarities between Leo's needs and Bascom's.

By her own account, Bascom has what she calls "normal, boring autism" and characterizes her support needs as "very middle of the road," since she acknowledges every autistic person needs some support, but some need very little. "There are people who need a lot more support, people who need twenty-four/seven, one-on-one support or even more than that," she said. "And then there are also a lot of autistic people who need less support than I do, who need support but can still live by themselves."

What Bascom means when she says she falls somewhere between the two is that she doesn't necessarily need someone tending to her constantly, but at the same time, she does need help navigating her daily tasks.

A native of New Hampshire, Bascom said when she first moved to DC she nearly died because of a lack of support. "I was only on my own for a couple of months, but I got very, very sick," she said. "Independent living in college was not something I would be able to do." Fortunately, Bascom had been speaking with Colton Callahan online, and he was already getting ready to move to Washington.

"I was looking to move to a city," he said. "Julia was looking for a roommate–slash–support person." Callahan's help is more a cumulative effect rather than assistance with one situation.

"Any one thing I'm helping her with, it's not the task itself that I need to help her with. It's all the little things that go into it," Callahan told me. For example, when Bascom is traveling, Callahan often prompts her to make sure she is eating and handling the sensory-processing issues that come with being in a large crowd. "It's not that traveling is a thing she needs help with so much as traveling has a bunch of little things that she needs a little bit of help or prompting with."

But even on a normal day, Bascom says she does not have the capacity to both hold a job and prepare meals for herself. "In theory, if I didn't have a job and everything else in my life was really easy, I could probably prepare two very simple meals for myself, like a sandwich and then something I heat up in the microwave," Bascom said. "That would take the whole day. And I can't do that. That's not a way to live a life. And I definitely can't do that if I'm working." As a result, when Bascom's workday is done, it's Callahan's responsibility to prompt her to do basic executive-functioning tasks, like showering or making dinner, which are especially difficult for Bascom at night.

"At the end of the day, I get stuck a lot," Bascom said, which she says is a reason she needs a support person. "You want to do something, but you can't make your body move."

To neurotypical people, this might just sound like typical end-of-the-day exhaustion when you don't want to cook and decide to order takeout instead. But as an autistic person, I know exactly what kind of sensation Bascom is discussing because I feel it as well. You literally are unable to move because you have spent your entire energy reserves trying to interact with people all day. It's the equivalent of running miles in a weighted bulletproof vest. Afterward, you have no more physical force to exert. It's why, as of right now, I don't live completely independently; I rent a room in a house in Washington that came furnished, and I hire a cleaning person to clean it once a month. There is no way I could have the executive-functioning capacity to do those tasks on top of everything else in my life.

Bascom and Callahan use a model for living called Shared Living, which is essentially where "an individual, a couple or a family in the community and a person with a disability choose to live together and share life's experiences," according to a brief prepared by the Arizona Developmental Disabilities Council for the National Association of State Directors of Developmental Disabilities Services. The intention behind these projects is to ensure disabled people "experience real community life," rather than one that is controlled by an outside body.

It is not entirely clear how many people use these types of arrangements. A 2020 report from the University of Minnesota's Residential Information Systems Project, which tracks the long-term services and supports that people with developmental disabilities receive, found that 5 percent of people with intellectual or developmental disabilities live in "host" or "foster family" homes, which has some overlap with shared living. However, it includes only those shared-living settings wherein the nondisabled person leases or owns the home. In addition, many shared-living arrangements like Bascom's and Callahan's are conducted outside the context of Medicaid, which means they aren't tracked.

Shared living allows autistic people to live fulfilling lives and not have to be overwhelmed by complex tasks. Bascom can go to Whole Foods, get a scone, and come back. But grocery shopping is not something she can do. Callahan told me that she will tell him that the employees at the grocery store will let her leave if she has a full cart. "And she's made that connection in her head. So, she'll just start putting things in the cart that we don't need, that we don't want. She just puts them in there, and then she'll buy them, and then she'll come home. And I was like, 'Why did you do this? Why would you do that?' And she was like, 'Because they let me leave if I do this.'" I completely understood what Callahan meant. Much of the world's rules are not explicitly written; people move through the world through general understanding and what they deem common sense. But if there aren't well-articulated rules to explain things, autistic people often have to work by mimicking others.

As a solution, Callahan goes grocery shopping for the both of them because he likes it, and this way Bascom does not need to be overwhelmed by interruptions to her daily routines. Callahan also keeps small problems from becoming big problems for Bascom even if he's not doing anything, as was the case when she needed to get her disabled-person Metro card renewed. Callahan went with her so she could find the right place.

"And it went perfectly smoothly. I went and I got my disabled card," Bascom said. Callahan said people might wonder why he went when he didn't do anything, but he explained that his presence alone can make a difference for Bascom. Without him, she likely would run into those small problems, like finding the door or knowing what to say to the front-desk person. Being able to navigate all of those small things allows her to accomplish the tasks with ease.

"I was there because we talked it out beforehand. We made a plan for where we're going and what we're doing," he said. "I was there in case things went wrong. But it's her errand, it's her life. So, she did it. I was there in case [I was] needed. And me being there was still integral. I was still necessary to the process."

I'm one of those autistic people who doesn't need daily support. I do not require a live-in support person, like Bascom, or need around-the-clock care, like Leo Rosa. Plenty of people have told me that they didn't realize I was autistic until I disclosed (those that know about autism usually guess, though), and that is probably because of the fact I live on my own. I pay my bills on time and ensure I am being a good neighbor. I don't have a home-care aide or my parents nearby. I don't require an around-the-clock aide or even a part-time aide. But this just means I have developed executive-functioning skills. That doesn't make me any less autistic than Bascom or Rosa, but it does mean that a lot of times, my capacity for competence is not called into question the same way Bascom or Leo Rosa's is.

"Just Set Me Free"

Despite the movement away from institutionalization and toward more community-based living arrangements like Julia Bascom's and Leo Rosa's, many autistic people still rely on some sort of government assistance when it comes to housing.

The major shift toward government-subsidized community living began in 1981 when Congress enacted 1915(c) of the Social Security Act, which effectively created Medicaid Home- and Community-Based Services (HCBS) waivers. These HCBS waivers were meant to provide services like in-home personal care and case management to elderly people as well as people with mental, physical, and intellectual disabilities as an alternative to institutionalization. In fact, the reason they are called waivers is that they allow states to "waive" the medical-assistance rules that govern institutional care to allow people to live in the community instead.

Despite the passage of this progressive-seeming legislation, initially, many people were slow to take advantage of this new service. However, two policy changes in the 1990s finally expedited the shift. First, the

Americans with Disabilities Act "instructed states to move to avoid the needless institutionalization of disabled people," as Pitney explained, cementing the idea that there are other, preferable living arrangements for that population.

Second, in 1999, Lois Curtis and Elaine Wilson, two developmentally disabled women with mental illness, filed a lawsuit when they were not transferred from the psychiatric unit of state-run Georgia Regional Hospital to a community setting. The Supreme Court ruled in *Olmstead v. L.C.* that unjustified segregation of people with disabilities violated Title II of the ADA and said that public entities must provide community-based services for people with disabilities when appropriate. This essentially meant that segregation of people with disabilities was now a form of discrimination and therefore was against the law.

"First, institutional placement of persons who can handle and benefit from community settings perpetuates unwarranted assumptions that persons so isolated are incapable or unworthy of participating in community life," Justice Ruth Bader Ginsburg wrote in her opinion for the Court. "Second, confinement in an institution severely diminishes the everyday life activities of individuals, including family relations, social contacts, work options, economic independence, educational advancement, and cultural enrichment."

These concurring policy moves led to a seismic shift toward community living. As Pitney wrote, in 1995, HCBS waivers accounted for only 30 percent of Medicaid's long-term support for people with developmental disabilities, but in 2012, 70 percent of spending went to HCBS waivers while only 30 percent went to institutions.

While a boon for the disabled community en masse, one problem is that HCBS waivers were not made with autistic people in mind. (Remember, autism did not appear in the *DSM* as a diagnosis separate from schizophrenia until 1980, the year before the creation of the waivers.)

As Julia Bascom explained, "They're based off of people with intellectual disabilities or people with physical disabilities." Often, this means that people without intellectual disabilities are ineligible for the waiver. (For instance, Bascom does not have a Medicaid waiver; she arranged her setup with Callahan between the two of them.) "And that means that people don't get the services that they need. Not everybody is lucky enough to have a friend who is willing to do this for many hours," she added.

In this same vein, autistic people's housing needs might vary more than the needs of people with other types of disabilities. For instance, some autistic people might not be able to handle fluorescent lights or would not be able to live near an airport because the noise could cause sensory overload, as Cal Montgomery said. On top of that, many autistic people do not drive, so they need to reside near public transportation. (That one I relate to; I tried learning how to drive multiple times when I was a teenager, and it was too much for me to process sensory-wise. It's a big reason I live in a metropolis with somewhat reliable public transportation.)

But states also have the capacity to cap HCBS waivers, which can create waitlists. One news report from 2017 found that Hamilton County, Ohio, where Cincinnati is located, had a backlog of 3,791 people, a 50 percent spike from four years before then. In 2019, New Mexico's backlog for its Developmental Disabilities Waiver Program was five thousand individuals over a six-year period and people waited roughly thirteen years to get services. A report from the Kaiser Family Foundation in 2019 found that in 2017, there were more than 707,000 people on waiting lists in forty-eight states. This backlog creates another dilemma for the autistic and otherwise disabled people:

Should they risk entering an institution or risk a waitlist for a waiver with little to no support in the meantime in hopes that the queue eventually clears up?

Still, despite the HCBS waivers' imperfections, there is evidence that they do overwhelmingly benefit autistic people. Since waivers are administered by state programs, they can vary wildly. One study published in *Health Services Research* in 2019 found that a generous waiver program significantly increased the odds that black autistic kids would have their needs met, with Maryland, North Dakota, Arkansas, Montana, Utah, South Carolina, Massachusetts, and Missouri having the most generous waivers.

A 2018 study from the *Journal of Applied Research for Intellectual Disabilities* found that families said that home- and community-based services autism waivers improved their overall family quality of life. Another 2018 report from *Medical Care* found that HCBS waivers were associated with reducing the number of unmet needs in autistic children, and more generous waivers were associated with fewer unmet needs. Incidentally, one aspect of the study was that the waivers appeared to disproportionately benefit children from higher-income families because autistic children from lower-income families already qualified for Medicaid, which tends to be more generous regarding autism than private insurance. Some may see this as a regressive policy, but this is a net positive both policy-wise and politically—children of wealthy people deserve to have adequate supports and services in their community as much as poor autistic children do, but more than that, wealthy people seeing the benefits of the program might be inclined to support the program and even improve on it.

In addition to better serving the disabled community, HCBS waivers can help save the government money. The National Council on Disability found that the average annual state expenditure for someone in an institution was $188,318; compare that to $42,486 for Medicaid HCBS. However, despite this financial reality and the decades-old ADA legislation supporting community living, government services are still biased toward institutions.

"Even when a person is on an [HCBS] Medicaid waiver, there are financial incentives to use group homes instead of supporting a person in their own home," Samantha Crane at ASAN said. Medicaid specifically prohibits paying for room and board for settings like apartments but does pay for them for facilities like hospitals and nursing homes. As Crane explained: "Technically the group home is supposed to separately bill for room and board, but because they are bundling that with other services they often can charge lower room and board fees than the average rent on an apartment."

In addition, as another report from the Kaiser Family Foundation said in 2013, states are required to cover benefits for nursing care, but most HCBS coverage is optional. And how states spend their money varies widely; in 2013, states spent anywhere between 21 to 78 percent of the money for Medicaid Long-Term Services and Supports on HCBS.

Crane added that there are also cultural attitudes that perpetuate the use of HCBS waivers to cover institutions: "Because group homes and institutions serve multiple people at once, they're seen as 'known quantities' to guardians and case managers, which means that it's often seen as easier to find a person a group home or institutional placement than it is to figure out the individualized services and supports that they may need in order to live in their own home."

"Your Sons and Your Daughters Are Beyond Your Command"

But HCBS, like everything in autism advocacy, has become a battleground between autistic self-advocates and their supporters versus (largely) parent advocates and their supporters, the latter of whom want what they believe is the safest possible option for their children. Many parents of autistic children also want to keep autistic people in some sort of isolated or segregated setting.

This fight even includes which types of settings can be considered "home- and community-based" when it comes to HCBS. In 2014, the Centers for Medicare and Medicaid Services released what was called the "settings rule," which was meant to determine which types of settings states could use Medicaid dollars to pay for HCBS, as well as define and describe the requirements for home- and community-based settings. The goal was to ensure that people with disabilities and seniors who received HCBS were truly integrated into the community. In addition, CMS also released a guidance document that outlined four types of group settings that have the effect of isolating individuals who receive HCBS waivers from their broader community:

- **Farmsteads, or residential farms:** Sending developmentally disabled people to rural areas and farms dates back to the 1850s, when the policy was to "protect" them. But the current incarnation can be traced to the creation of Bittersweet Farms, which was established in 1983 near Toledo, Ohio. Bernard Rimland even said in a blurb of a book examining Bittersweet that the format could be a future setting for his own son Mark. But many advocates see them as a form of segregation because they are run without the input of disabled people. Autistic self-advocate Kit Mead wrote in 2016 that they were akin to the colony model of institutions for developmentally disabled people during the late nineteenth and early twentieth centuries. "Often these farm arrangements are custodial, and fail to move people out of the isolated setting into the community with real jobs that pay real wages, not piecemeal subminimum wages," they wrote.

- **Gated or secured community for people with disabilities:** Defined as consisting "primarily of people with disabilities and the staff that work with them." The reason these places were considered isolating was that people who received services there did not leave the community to interact with the broader community, thus sequestering them.
- **Residential schools:** Facilities that provide education and residence. In 2012, the *New York Times* covered the draft of a report that said that residential schools run by New York's Department of Education did not track abuse claims. The investigative news website ProPublica also reported in 2015 on residential schools' use of restraints and how alternative methods were largely self-imposed, rather than mandated.
- **Settings that are located near each other and are operated and controlled by the same group:** These included places like group homes and intermediate-care facilities, which provide long-term care for disabled people. It also included group homes that are on the same site or close to each other. The guidance said that the shared programming and staff limited residents' ability to interact with the larger community.

Bascom said the settings rule came about because far too often, people with disabilities were put on waiting lists, which meant they would waive their right to institutional care while also going years without services they needed.

"They finally get their alleged HCBS, and it would be indistinguishable from an institution. A group home would have the same rules," she said.

Incidentally, some parent advocates support the use of HCBS waivers to be used toward more institutional-style living like the options just explored. One such parent is Amy Lutz, who would later help start the National Council on Severe Autism and who has an autistic son named Jonah. In an op-ed for *Spectrum News,* Lutz went after the American Civil Liberties Union for siding against these congregate settings and said it showed "a troubling display of paternalism." She even compared the ACLU to the doctors who founded the first institutions in the nineteenth century; they "believed they knew what was best for people with disabilities too," she wrote.

"Some adults with disabilities and their families are attracted to large, disability-specific communities because of their social benefits. But others—like Jonah—need this option, as they require more support than can be safely and consistently delivered in dispersed settings," Lutz wrote. She added that one of her fears was that when she and her husband were no longer able to care for Jonah, he would end up in his own apartment with an aide who wasn't adequately paid and thus didn't properly care for him.

Similarly, in 2016, Paul Solotaroff, the father of an autistic young man, expressed his frustration with the rule in a longform piece in *Rolling Stone,* which said CMS's rule "launched a strike on intentional communities," a term many supporters of these types of settings use. It also quoted Jill Escher, who was then president of the Autism Society San Francisco Bay Area (and who would later become a founding member of NCSA), who accused Medicaid of making the rule to cut financial costs. "They took a look at their budget costs and said, 'To hell with them. Let's cut their spending now.'" Solotaroff envisioned his son, who was seventeen at the time, being able to move to a farm in Massachusetts that had gotten around the final rule by not having residents live in group homes but rather share homes with provider families.

Public policy has also begun to move in favor of parent advocates like Lutz and Solotaroff. In 2017, CMS under the Trump administration gave states until 2022 to show they are complying with the new rule when they were initially meant to comply in 2019. In 2020, CMS again extended the compliance deadline until March 17, 2023, amid the coronavirus pandemic.

Then in 2019, CMS released new guidance on the settings rule. While it kept the rule in place, it changed the criteria—which it claimed it streamlined—for settings that isolate and removed the list of specific settings that isolate. Most notably, it also said, "Settings located in rural areas are not automatically presumed to have qualities of an institution, and more specifically, are not considered by CMS as automatically isolating to HCBS beneficiaries." In turn, if states determine that gated communities and farmsteads met the criteria for a home- and community-based setting, they could be eligible for Medicaid HCBS funding. The new guidelines said that states should only submit a specific setting to CMS for scrutiny if it had qualities akin to institutions. Director Seema Verma called the previous guidance "too prescriptive" and said it "unfairly singled out certain settings," which Verma said worried beneficiaries, families, and providers. Still, the new guidance kept the settings rule in place.

"Even well-intentioned policies from Washington often lack the flexibility needed to work for every state, community, setting, or family," CMS director Seema Verma said at the time in a press release. "We believe our revised guidance strikes the appropriate balance to protect individual choice while maintaining the integrity of home- and community-based funding."

NCSA and other groups hailed the clarification, calling it a victory for "choice." Not surprisingly, many autistic people and self-advocates did not hail it, though they were relieved the settings rule stayed in place. There is an understandable impulse behind these anxieties and fears. These parents do want what's best for their kids and worry about what happens when they are no longer capable of caring for them. They might see the original restrictions against those settings as compromising their sense of security.

Similarly, Alison Singer, who is treasurer of the NCSA and the president of the Autism Science Foundation, told the Interagency Autism Coordinating Committee that intentional autistic living communities —which supporters describe as supervised housing for developmentally disabled autistic people with support staff and programming— are nothing like the institutions of old. (Singer, some readers might remember, was the spokesman from Autism Speaks who left amid the imbroglio about vaccines.) Lutz also quoted David Mandell, the professor at Penn who surveyed a state hospital (and who has coauthored many of the studies in this chapter), who told her that there is no data proving that living in apartments is safer than living in an intentional community.

But Julia Bascom disagrees. Many of these places are essentially neo-institutions that are "shot-for-shot recreations of the first institutions," she said.

"I think if someone thinks an intentional community is so great, they should go live in one. But it's usually not the disabled people asking for them," Bascom told me. Bascom's colleague Sam Crane has also said that they are still isolating because they are "designed so that people have no need to leave the campus on a daily basis except on group trips."

Bascom said she understands the desire of parents to make sure that their kids are protected, which leads to the desire to build a place for children to be kept safe. "But the reality is the thing that's going to keep him safe is those relationships with other people."

Similarly, Montgomery says it is easy to give the illusion of choice in an institution, but self-determination is the most important aspect.

"One of the things that I learned when I worked in group homes is how incredibly easy it is to create an environment that looks like it is all full of choice and happiness and whatever," he said, when in reality it's the people who work there who are in full control.

Montgomery's priority for any type of living situation is independence. But the problem is that any time there is a new system put in place for people with developmental disabilities, "the same old people are working it," and those same people will just take over and take control like they did in the past. Instead, Montgomery said, there needs to be a focus on finding what best promotes self-determination. "But I don't think that's gonna just do it. I think if we got everybody into, what is now considered HCBS, I think for most people in the system, they'd still be institutionalized," he said. "There'd be smaller places, it'd be places that are easier to address. And that are in many ways better, in most cases, but I don't think it solves it. So, I'm still struggling with what we do."

Montgomery said the places would be smaller and probably better, but it wouldn't solve the problem of lacking autonomy and independence or being free of the restrictions of institutions. Often, he said, people who were institutionalized were moved presumably "to the community" but were given "wholly inadequate services." Similarly, people who live in big developmental centers are often "in the community," but their services are still restrictive and coercive in nature. In other words, whenever a new solution is created for intellectually and developmentally disabled people, including autistic people, the same old problems plague them.

Some states have additional services to assist autistic and developmentally disabled people outside of institutions. For example, the Rosa family's insurance covers in-home services. California's regional centers also fund respite for people like Leo. The regional centers fund 80 percent of Leo's camp in a location called Via West in Silicon Valley, and the people have one-on-one or one-on-three support for the entirety of the camp. But the Rosas are lucky they live in California, because how good the services autistic people receive can be entirely contingent on the state where they live.

"I Am the One Thing in Life I Can Control"

The debate between some parent advocates and self-advocates is not limited to where autistic people live but also how they live, particularly on the question of guardianship. While American guardianship in its current incarnation can trace its ancestry back to late fourteenth-century British common law, many disability rights activists and promoters of independence argue that it limits disabled people's independence. One of the defining traits of independence as American and Western liberalism defines it is the right to have options and decisions. But many advocates object to guardianship because it means that someone else can control their finances, where they choose to come and go, whether they can vote, and even whether they can get married.

In response, many states have adopted supported decision-making as an alternative to guardianship. In guardianship, disabled people often have little control over their lives without the consent of their guardian. Supported decision-making allows people to make their own decisions about their health care, residence, and finances but with the assistance of people they designate as their supporters. Still, at the end of the day, it is disabled people who have the final say.

"Guardianship involves substituted decision-making and the right of the person to make decisions is voided," said Michael Kendrick, who is the director for Supported Decision-Making Initiatives at the Center for Public Representation, a disability rights advocacy organization. "At least in a legal sense, [disabled people] lose their rights. Whereas in supported decision-making, they preserve their rights in their entirety. And in addition, they get assistance with their decision-making."

Furthermore, there seems to be bipartisan consensus on supported decision-making. The first two states to pass legislation for supported decision-making were deeply Republican Texas in 2015 followed by Delaware, an overwhelmingly Democratic state, in 2016. Other states that have laws on their books include Indiana, Nevada, Alaska, Wisconsin, North Dakota, and Rhode Island, along with the District of Columbia. Georgia and Massachusetts have both conducted pilot programs, and other states, like California, offer it without a formally enshrined statute. ASAN's Samantha Crane helped with the Texas legislation, Bascom said.

"We're talking about fundamental things like life, liberty, and the pursuit of happiness and your ability to be in charge of your life," Bascom said. "So, sometimes in blue states, you see more resistance because people are more used to this mindset of 'We have to take care of people,' and care means control for a lot of people."

Anna Krieger, a lawyer with the center who helped people set up their agreements in Massachusetts, notes that while Maine did not have a comprehensive piece of legislation, it did something equally important—it inserted into its guardianship a statute that supported decision-making had to be considered first.

"They weren't able to get that bigger bill passed, but they were able to get that key piece of information in," she told me. "So, with that as the foundation, I think that will help them move towards supported decision-making as the default option and then guardianship as the second choice."

But Crane notes that while supported decision-making is good on a state-by-state level, it must be accompanied by Congress passing federal legislation, such as reauthorizing Money Follows the Person, which provides financial assistance for people to leave nursing homes and was first signed by President George W. Bush. But since its funding lapsed, it has regularly received only short-term extensions, which has caused a drop in people transitioning out of institutions.

"If you don't have funding available to you, then you have no meaningful options to choose among. But once you do have those options, [supported decision-making] is really helpful," Crane told me, noting that community integration means allowing people to set their own daily routines and schedules, which supported decision-making can make a reality. "Whereas a person under guardianship may not be able to make their own choices about where to live, how to spend their discretionary income, or who to form relationships with—making community integration much more difficult."

At the core of so many of these debates about autism and independent living is the capacity for autistic people to make their own choices and have their own sense of autonomy within their communities, regardless of what that looks like.

I have a semi-autonomous life. Even though I rent a room in another person's home, I can come and go as I please; I don't have people setting hard hours or curfews for me. I can go out when I want and stay a little late with friends if I so desire; I can stay home and sleep in as long as I want on the weekends. Nobody forces me to go to work or come home at a certain hour, and I get to buy my own groceries.

Living independently took trial and error. In the weeks before I left to go intern at the White House, my mom taught me how to clean a bathroom and do laundry. When I arrived in DC as a White House intern, my dad helped me find the nearest grocery store, which turned out to be a bodega. Given Washington's sweltering humid summers, I needed to live near a Metro station and a store to get food. My mom insisted I get a cleaning person to clean my quarters once or twice a month, and I realized life could be easier if I had groceries delivered to my place.

I doubt in this phase of my life that I would be ready or even capable to move out on my own and handle all executive-functioning responsibilities that come with living alone. And that is okay. I doubt I would have my career if I needed to constantly tend to home-care issues. In the past, I used to hate when people prodded me to get more assistance because I felt like they were telling me how to live my life. Nowadays, I recognize that the support systems I do have in place, like my cleaning person, are essential for me to continue living my life.

Some may say that I am freer than Bascom because I do not need a support person. But in the same way that a wheelchair is not something that binds a person who cannot walk but rather gives them mobility, or how the glasses Callahan wears allow him to see, having a support worker actually gives Bascom freedom rather than limits it.

This isn't to say that all of my decisions have been correct. God knows I do not always eat the healthy option (after a long day at work and especially after writing, it is so easy to get takeout instead of cooking. And the Chinese food carry-out is literally across the street). In my independence, I have spent money frivolously on things I don't need. I've trusted people I shouldn't have. I once went a whole three months wearing the same undershirt beneath my dress shirt because I didn't realize you needed to constantly wash them. One time when I got back the wrong shirt from the dry cleaner's, I didn't return it for months, even though I knew I needed to. I wanted to, but it just felt so impossible to do and the task kept falling by the wayside.

You may be thinking, *But this is all part of adulthood; being human means you will inevitably screw up.* But while yes, the right to grow up also includes the right to screw up, only able-bodied people have this luxury. For autistic and other disabled people, every bad decision becomes a referendum on your right to live independently. Despite jokes about "adulting," nobody is going to force my neurotypical friends to move back in with their parents if they forget to do their laundry. If they need a nanny to care for their kids, a maid to clean their house, or a gig worker to do daily tasks, they won't be seen as incomplete humans. Autistic people's use of these services is no more of a measure of their incompleteness as humans than people who use wheelchairs are failures because they cannot walk.

To be disabled is to constantly fear that any bad decision you make will cost you your autonomy, particularly when there's a historical precedent for institutionalization. It makes your freedom all the more precarious. As Ruti Regan, an autistic rabbi and activist, has written, "The risk of failure is often higher than it is for people without disabilities."

But by the same token, Regan writes that when disabled people are not allowed to fail, we are not allowed to succeed. "Because for all people, success rests on a lot of failed attempts. And because disability typically involves uncertain abilities, we usually need to make a lot more failed attempts than nondisabled people as we figure it out."

It's a sentiment Callahan echoes and tries to respect in his partnership with Julia: "I get to make dumb choices and have no one check them," he told me. "If Julia wants to be like, 'I don't want to brush my teeth ever,' and then gets a million cavities, that's her choice. She can do it."

For those who have a limited perspective of what autistic people can do, it might be difficult to process that Bascom is a major figure in the disability rights movement. She is someone who has steered ASAN during a time when her community's rights are at risk. Her life shows the futility of functioning labels; each autistic person has the capacity to function on a really high level along with trying to manage the difficulties the world constantly throws at us.

In 2018, Bascom delivered a speech at the United Nations in New York City that received a positive reception. "I was really glad that I was able to do it. But what people don't know is that I had a total meltdown trying to get through security at the UN. It was awful." At the time, Callahan wasn't there; she had a different support person. Bascom explained that during the meltdown she insulted somebody. By the end of it, her support person tried to get her out of there, and Bascom didn't speak to the point that her talk had to be rescheduled.

"The same person did both of those things—gave this speech that was very well received and also needed a lot of support in order to be able to do it," Bascom said. "I think that people often think that if someone needs support, that means that they can't do things. And that if someone is doing stuff, that means they don't need support."

Still, there is this expectation for autistic people that life, liberty, and the pursuit of happiness are not rights but a diploma that they must earn. Around the same time as this UN speech, Bascom met a mother who was told her ten-year-old son needed to get his meltdowns "under control" or he wouldn't have a normal life.

"And I was like, 'I've had two meltdowns in the last month.' Absolutely, we need to support your son to find a better way to communicate before he gets so upset. No one likes having a meltdown," Bascom said. "But also, he will keep having meltdowns for the rest of his life, and he can still have a good life. And so it's hard to balance the consequences of talking about support needs."

Bascom wants people to know that it is okay for autistic people to need a lot of support and that it is okay for them not to live on their own. "You can still have a really good life. And the key to that part has to do with what support and services our government will provide and pay for, one hundred percent in order to access a lot of this. Or else you have to be very lucky and have an actual support network that can replicate that for you." But having those supports comes down to whether the government will pay for them, and if it doesn't, then people have to be either incredibly rich or lucky.

Bascom told me her life is fulfilling because she is able to do things that make her happy, like spending time with her cat or doing a low-key activity like watching a movie or going to Whole Foods to get a scone. Too many times, she said, autistic people are told that if they like things that deviate from the norm, those things don't count as fulfilling, particularly if they have significant support needs.

"But a good, happy life looks like someone who's in charge of their life and who's connected to their community and who's doing things that are important to them, not someone who is working a nine-to-five job and living on their own and meeting all the milestones," she said.

In summary, Bascom, the Rosas, Montgomery, and so many others demonstrate the many possibilities for autistic independence.

5

"Somebody Get Me a Doctor"

HEALTH CARE

Autism's years of being perceived as everything from a symptom of childhood schizophrenia to a condition caused by unloving mothers has led to people fundamentally misunderstand where it fits on the larger continuum of public health. It also means there has been both a booming industry of pseudo-cures and a misperception of what autistic people's health needs are—should the focus be their autism or on helping them have holistically healthy lives?

The question of what to do with disabled people, including autistic people, has been debated for decades. For years, disability was seen through the lens of the so-called medical model, wherein disability itself was the problem. By contrast, disability rights activists adopted the social model, which states that the problem isn't disability but rather that society is not accessible to disabled people. In this view, the world needs to shift its paradigm about disability to become more welcoming to them. This view doesn't diminish the needs of disabled people or the fact that disability can have complications. Rather, it recognizes the needs that disabled people have and works to give them services so they can live more fulfilling lives. At its core, the social model advances the civil rights of people who have certain developmental disadvantages but are not inferior because of them. They just need assistance to ensure equity.

There are some similarities between the social model of disability and neurodiversity, but as disability rights advocate Aiyana Bailin wrote, "The neurodiversity approach is primarily a call to include and respect people whose brains work in atypical ways, regardless of their level of disability," and it requires challenging assumptions of what's normal. Bailin means that impairments that autistic people face "don't change a person's right to dignity, to privacy, and to as much self-determination as possible, whether that means choosing their career or choosing their clothes."

These contrasting models were on display in the early days of Ruth Christ Sullivan's NSAC. As Steve Silberman's *NeuroTribes* describes, Bernard Rimland was focused on promoting treatments to make autistic people more "socially acceptable," whereas Sullivan often asked for a show of hands at meetings to see whether people wanted to focus more on finding a cure for autism or lobbying for more services. As Sullivan said, "Nearly all parents' hands went up for services." But decades after Sullivan's and Rimland's fights, there are still pushes to "fix" autistic people, and many still see the large number of autistic people as cause for alarm.

In 2017, his first year as president, Donald Trump held a roundtable with educators and asked one principal, "So, what's going on with autism? When you look at the tremendous increase, it's really—it's such an incredible—it's really a horrible thing to watch, the tremendous amount of increase."

That same year, I was at an event on Capitol Hill writing a profile for *Roll Call* about Senator Cory Booker as he tried to beef up his progressive bona fides before he launched an ultimately doomed 2020 presidential campaign. I had already interacted with the New Jersey Democrat once before and had written a few articles about him. At times, his staff didn't like my coverage of him, but I always tried to be fair, as I do with any Democrat or Republican that I cover.

I wanted to catch Booker in person and ask him a few questions, but first I had to listen to his lecture about the economy and the Federal Reserve. I used to attend countless events like these on Capitol Hill, usually in a reception room in one of the Senate's office buildings. The guests were economists, lobbyists, activists, and even some former members of Congress who just couldn't seem to quit Washington even after they retired. I started my journalism career covering economics, so I do find the subject fascinating, but occasionally I am reminded of what President Lyndon Johnson said on the topic: giving a speech about economics is like pissing down your leg—it feels hot to you but not to anyone else.

In his speech, Booker discussed how corporate America had abandoned cities like Newark, New Jersey (where he was mayor before elected to Senate), and how that led to a spike of illnesses among children. "Cancer rates, autism rates, asthma rates in communities like mine, undermining the potential of my children, are all a result of the perversion of the free market," he said.

Booker's words equating autism with cancer and asthma stuck with me. Whatever my thoughts about him, his exuberance and desire to inject what he called "unreasonable, irrational, impractical love" into American politics led me to believe he was speaking with the best of intentions. And Booker, a Stanford graduate and Rhodes Scholar, has a sharp mind. But Booker, like so many other well-meaning and intelligent policymakers, saw autism as the bad fruit born from a rotten tree of corporate greed. But while everyone wants asthma and cancer eliminated, not everyone wants autism eradicated, nor is there much evidence that it can be cured.

Tunnel Vision: The Misguided Search for a Cure

The focus on curing autism goes back to its initial recognition as a disorder in the United States. In the years following the publication of Leo Kanner's research, autistic people were subjected to a variety of attempted treatments, from electroconvulsive therapy combined with subcoma insulin shocks and the drug metrazol, which caused convulsions, to psychotropic drugs like LSD. The idea that autism was the fault of parents caused significant damage to the way health-care professionals understood autistic people, setting a precedent in which doctors attempt to "cure" their symptoms rather than focusing on their overall well-being. As mentioned before, part of Bruno Bettelheim's rationale for removing children from their parents was "thawing" them out from their icy home lives; he claimed that 92 percent of speaking children had "good" or "fair" outcomes and that the children classified as "good" could be considered "cured," though his methods and claims were questionable, at best.

The changed and broadened diagnostic criteria for autism in the 1980s and 1990s naturally meant that more people would be diagnosed as having it. Unfortunately, that led to many policymakers bemoaning an "epidemic." Using the language of disease to describe autism led to a panicked fixation on a cure that still exists to this day.

Though Autism Speaks officially removed the word *cure* from its mission statement in 2016, when the organization was founded in 2005, the goal was to eradicate autism. And Bob and Suzanne Wright's mission was not unusual; many other parents and family-organized nonprofits have worked to find a cure or treatment for autism, including Bernard Rimland's Autism Research Institute—with its Defeat Autism Now! network—and Cure Autism Now (CAN), founded by parent advocates Jonathan Shestack and Portia Iversen, which later merged with Autism Speaks.

Before he died, in 2006, Bernard Rimland specifically said it was up to parents to ensure their children's cure because doctors had an incentive to treat autism with drugs instead of curing them, which he believed led them to dismiss his chelation, nutritional therapies, and diets as quackery.

"So once again, it is up to parents to lead the revolution—and we have not one but two battles to fight," he wrote in *Autism Advocate,* the magazine for the society he founded, noting the first battle was to provide "safe, effective, biologically based treatments" for children and the second was to "create a safer and saner world for all of our children, and generations to come" by preventing exposures to PCBs, lead, mercury, pesticides, food dyes, food additives, and "nutrient-stripped food," all of which he saw as "damaging the bodies and minds of generations."

Though many of these organizations were operating according to the popular rhetoric of the time, treating autism as something that must be cured can have dangerous consequences. Most notably, there are a slew of cottage industries offering supposed "cures" to parents of autistic children, each one varying in its level of dubiousness. For instance, while Andrew Wakefield's ideas about vaccines have been discredited, his impact on clinicians remains. He popularized—though did not create—the idea of "leaky gut syndrome," wherein vaccine particles prevented the breakdown of certain foods, like wheat and dairy, which supposedly passed through the walls of the gut, made their way into the brain, and caused autism (if this sounds absurd, it's because it is). Similarly, Bernard Rimland's Defeat Autism Now! warned against gluten and casein (a protein found in milk) in autistic children's diets.

But contemporary science refutes these sorts of diet-based treatment plans. In one study of sixty-six autistic children done at the University of Warsaw in Poland, researchers placed half the subjects on a gluten-free diet and the other half on a normal diet with gluten in at least one meal. After six months, the scholars definitively concluded there was no evidence that gluten consumption affected the symptoms of autistic children.

In recent years, some parents have turned to an even darker "solution" for "fixing" their children: forcing them to drink chlorine dioxide, which is essentially bleach. Jim Humble, a former Scientologist, named the compound the Miracle Mineral Solution, or MMS, and sold it as a cure for a litany of different illnesses like AIDS, cancer, and diabetes. Eventually, Kerri Rivera, who was a real estate agent in Illinois with no medical training, began to experiment with MMS on her autistic son and sell it to others.

The concept is horrifying, yet the practice is so widespread that the U.S. Food and Drug Administration had to publish a warning saying that MMS distributors were making "false—and dangerous—claims" and that people should avoid their phony solutions. It is certainly disturbing that parents would willingly give their children a dangerous chemical as a means to an end. And the list of bunk treatments goes on, among them placing kids in hyperbaric chambers, where people breathe pure oxygen in a highly pressurized environment, and chelation therapy, which removes mercury from the body (based on the mistaken idea that mercury in vaccines causes autism). It should be noted that there is scant evidence for the effectiveness of these treatments, and there is evidence they can be harmful. What's even more disturbing than these treatments themselves is the logic at their core—that it is better to gamble with children's health by giving them pseudoscientific treatments than to let them be autistic.

When it comes to treatments for autism in the United States, perhaps the most intensely debated regimen is applied behavioral analysis (ABA). ABA is a type of therapy meant to help people improve basic skills such as communication, hygiene, and fine-motor dexterity using conditioning. It is so hotly contested that there is even a debate about whether health insurance should cover it.

The University of California, Los Angeles, psychologist and professor Ole Ivar Lovaas pioneered the practice, in which he encouraged breaking down bigger skills into smaller, manageable increments. But Lovaas's form of ABA was harsh; a 1965 profile of his treatment in *Life* magazine titled "Screams, Slaps and Love" featured researchers rewarding good behavior with food and affirmation and punishing bad behavior by slapping their autistic patients, screaming at them, and, in the most horrific example, administering electric jolts. Lovaas referred to autistic children as "little monsters," in a 1974 *Psychology Today* interview and derided the way autistic children "rock themselves back and forth or spin around in a circle," never considering that this could be a form of self-regulation.

Eventually, Lovaas removed the punitive aversive methods from his program, and since then, ABA has come to be viewed as the "gold standard" for autism treatment in the United States. As of 2017, forty-six states and the District of Columbia require coverage of autism services, and many of them require insurers to cover ABA, according to the National Conference of State Legislatures.

But many autistic advocates say ABA focuses on eradicating autistic behavior. Before Hari Srinivasan began using a communication device to speak, he went through ABA. But Srinivasan said it did not succeed, since it was hard for him to give consistent receptive responses, given that he deals with sensory dysregulation (difficulty with sensory processing) and oral-motor apraxia (the inability to sequence oral movements of speech structure for nonspeech tasks). His body schema (which is essentially awareness of the orientation of one's body) is poor, and he has obsessive-compulsive disorder and minimal fine-motor skills.

"It was groundhog day as I was stuck in the same lessons for years which gets boring real fast," he told me in an e-mail. "It is unfortunate that a lot of gatekeeping happened as my oral-motor ability was being looked at rather than cognitive ability so I was always pushed towards special ed classrooms, which were placeholders."

Still, there are some people, even autistic people, who swear by ABA, saying that the practice varies depending on who administers it. Amy Gravino, who is autistic, was inspired to pursue a master's degree in ABA and now counsels students at Rutgers University in New Jersey.

"I very much consider myself a moderate. In every other circle in the world, I'm a liberal," she said. "But you put me in the autistic community, and I look like a raging conservative next to these people. It's freaky."

Gravino said she feels defensive when other autistic people call her a traitor. "I'm not here to invalidate what anybody feels or to dismiss what anybody has gone through or to diminish that," she said. "I pursued my degree because I believed that I could bring a voice to the ABA world that was not perhaps represented in that world."

The debate about applied behavioral analysis ultimately comes down to how autistic people should be treated. Plenty of parents enroll their children in ABA, hoping to give them some modicum of hope for a normal life. But the parameters of what is considered a "normal" life are often defined by people who are not autistic. While there certainly should be a focus on stopping behaviors that cause harm to autistic people or that violate the personal space and safety of others, that impulse cannot come at the expense of things that make autistic people themselves. They must not be forced to suppress what makes them who they are.

However, some autistic people do want to be cured. As Benjamin Alexander, an autistic person who cannot speak, told *Spectrum News* in an interview, for him autism is a "living hell" and "I want to be cured just like a cancer patient wants to be free of disease."

At this point, it is important to state that fighting for neurodiversity or celebrating disability does not mean that autistic people think that their condition is all good. They still recognize that it is a disability.

Jill Escher, one of the cofounders of the National Council on Severe Autism, has two autistic sons who cannot speak, and she believes that autism is a "severe neurodevelopmental disability." She said neurodiversity advocates trivialize the condition or "cherrypick naive, feel-good stories that portray autism falsely instead of grappling with the reality." Escher added that if her kid is having a meltdown, taking his clothes off, or screaming, then "I want people to appreciate that his behavior comes out of a difference in his brain wiring. But do I think his behavior and wiring is natural? Absolutely not."

But embracing autism or accepting autistic people for who they are does not mean ignoring the legitimate challenges. Far from it. It simply means acknowledging that autistic people and all neurodivergent people deserve the same civil rights as others, which advocates like Sam Crane at ASAN have articulated. Often, they are the ones who want to include it in the larger movement for disability rights and request more accommodations. Many of them recognize that some autistic people have more impairments than others and want to find ways to help autistic people with comorbidities like epilepsy and gastrointestinal issues. Embracing autistic people and acknowledging their needs are not mutually exclusive ideas; they are complementary.

For the most part, the increase in diagnoses has given autistic people something important: a community. Nancy Bagatell, the director of the University of North Carolina's behavioral science and occupational therapy department, wrote that this community has fostered a discourse where autistic people are "considered worthy individuals in and of themselves, not people who need to be cured, altered, or isolated from the world" and that autism is "seen as a fundamental part of who they are, not just something that they have; that is, if their autism were eliminated, they would not be the same person."

Acceptance of autism doesn't mean a denial of those impairments any more than accepting other disabilities means denying their impairments. One of the best descriptions of this dichotomy came from autistic author and advocate John Elder Robison.

"Autism is a unique condition in medicine because it confers powerful disability and really extraordinary exceptionality," he told me back in 2015. "Our duty in autism is not to cure but to relieve suffering and to maximize each person's potential."

"I Just Didn't Know That You Had to Play Social Games with Doctors"

I arrived in Allison Park, Pennsylvania, a suburb of Pittsburgh, in February 2020, when snow layered the Allegheny Mountains. I was there to visit thirty-two-year-old Lydia Wayman. The week before, she had moved back in with her parents because her live-in nurse is retiring, though she still planned to continue working part-time to help with ordering supplies, which could be overwhelming for Lydia (prior to the COVID-19 pandemic, the nurse came two to three times a week). Lydia, Lydia's mom, Nancy (who goes by the nickname "Brisky"), and I sat in their living room to talk about how difficult it was to have her needs met as an autistic person with chronic health issues. Lydia has autonomic dysfunction, which contributes to digestive issues; a vascular condition that causes her skin to turn red and is excruciatingly painful; and debilitating migraines.

Brisky, who has short silver hair, now helps with Lydia's care overnight, which is her worst time because her erythromelalgia flares prevent Lydia from getting up and handling some things herself, and standing and moving aggravates it. Nighttime is also when she has issues with her blood sugar, pump alarms, and autonomic dysfunction; her skin can take hours to cool down, and the ice packs in her cooling system need to be changed. A separate nurse comes by twice a week to draw labs and change dressings while her mom works as a personal care assistant.

The way Lydia's care is paid for further illustrates how navigating the health system can be tremendously frustrating for autistic people. Medicaid pays for her overnight nurse and her personal care assistant for a set number of hours per week. But her insurance, specifically Medicare, pays for the nurse who does her lab work and changes dressings.

As we sat in her house, Lydia explained that she kept a backpack next to her that contained her IV. Lydia also has an ostomy bag and an insulin pump to manage her type 1 diabetes, and prior to that, she had a feeding tube for seven years.

Lydia has always been autistic and has had diabetes since she was three, but her medical ordeals in the past decade show how ill-prepared the health-care system is to handle the needs of autistic people. Lydia struggled to navigate arcane medical processes and eventually landed in a nursing home for a little over a year.

It all started in January 2012, when she began vomiting profusely.

"I guess it started with my stomach, but my digestive tract is almost paralyzed. It doesn't really work," she told me. "So things just would sit in my stomach, basically, and not make it through, and I'd just throw up."

Wayman went to the emergency room four times between January and April of that year to be treated for dehydration. In March, she was diagnosed with gastroparesis, which means her stomach is incapable of emptying food. That November, Wayman discovered she had developed autonomic dysfunction, which prevents her from sweating.

There is also her skin. When we met at her home, Lydia was wearing makeup, so her complexion was relatively pale and similar to that of her mother's. But Lydia told me that her blood vessels stay dilated and showed me a picture on her tablet in which her skin was glaring red. Her ears, back, hands, and shoulders are also visibly red. The name for this condition is erythromelalgia, also known as "man on fire syndrome."

When Lydia first got sick, back in 2012, she was living in an apartment on her own thanks to aid from Pennsylvania's Adult Autism Medicaid Waiver. But once she got sicker, her illness prevented her from participating in the requisite goal-building—such as grocery shopping, planning and using a schedule on a computer, attending church and Bible studies, and learning social cues—that she needed in order to continue receiving aid.

"I got to participate in discussions about what the goals should be. I just didn't always agree with how they got implemented," Lydia told me. For instance, she was supposed to learn about social cues through watching television shows like *Friends,* which involved social scenarios she would likely never participate in. This led to Lydia losing her waiver. She applied for a physical disability waiver instead but was told that since her "primary" diagnosis was autism, she had to apply for the very waiver that she had already lost. Lydia said she fell into a "crack [that] isn't supposed to exist" within the system.

That fall through the crack landed Lydia in a nursing home for fifteen months. Lydia said she ended up there because other options didn't really exist where she lived in Pennsylvania; people needed to handle their own housing in the community and get waiver services or go to a facility. She said that housing supports existed only for people with significant mental illness or intellectual disabilities. Brisky said the facility Lydia went to was one of the only ones willing to accept her daughter.

"They were one of the few, and they did the best they could, but that's not really what they were set up for," Brisky said of the institution where Lydia lived. "There is a place for people like Lydia that have medical conditions and they're young people and they have chronic medical conditions, and they need support. I don't know where it is." In addition, prior to Lydia's time at the nursing home, her parents were told not to get too involved in the process because they were supporting Lydia's "bad" habits and that Lydia's living with them would encourage her to stay "sick." It was overwhelming for her parents, since they wanted to advocate for Lydia in a system where she needed help.

"I just didn't know that you had to play social games with doctors basically," Lydia said.

Eventually she left the nursing home, and after applying for the waiver a fourth time, she finally qualified. A woman from her county's center for independent living helped her with the process.

"So, every single phone call, form, she was with me through the whole entire process and she also meticulously followed up with the state," Lydia added.

Before her time in the nursing home, Lydia faced another obstacle—she said she had a primary care physician who doubted the severity of her illness. "She concluded that I was creating my symptoms to get attention, but I didn't realize I was giving her that impression," Lydia told me. Ironically, Lydia might not have been taken seriously because, as an autistic person, she was trying to follow the cues that neurotypical people gave her. "People always told me to stay positive, and I took that at face value." Wayman was also told she was "too happy to be sick." (In an almost cruel twist, once Wayman was in the nursing home and under supervision, it was clear she wasn't faking anything.)

Subsequently, her physician contacted her gastrointestinal doctor to warn him that Lydia was obsessed with getting sick; she said Lydia knew too much medical information.

"For me, knowledge was a way to cope with the fear and uncertainty my doctors didn't believe I was feeling," she said.

To be disabled, chronically ill, or autistic is to be constantly aware of your mind and your body and how the world limits them. Because of a lack of understanding from doctors, knowledge becomes a protective shield. Lydia said other autistic people she knows, including two of her best friends, have developed chronic medical conditions and agree that knowledge is the best way to protect themselves.

"Because we [autistic people] rely so much on familiarity or routine that when something so fundamental changes, you're so lost, and you've got to process it," she told me. "Every little thing you do is now different. Every way that you learn to expect the world to be and your day to go and your causes and effects to happen is different. And so, you have to learn to cope with life in a whole new way, and it takes away your focus for a while."

Lydia later clarified that her experience was more complicated than just having been forced to deal with bad or mean doctors. It can be challenging for autistic people to communicate with their doctors successfully in a system that's built with neurotypical patients in mind. She said that while her symptoms didn't contradict each other, they appeared to be disconnected. Furthermore, Lydia's sensory issues made it hard for her to articulate *how* she felt and communication disorders made it hard to explain *what* she felt. Once she had more testing, everything began to make sense.

"My doctors did misunderstand my behavior, but I'm sure I was missing social cues that they were doubting me," she said. "They knew I was autistic, so for me, the idea that my autistic traits could affect my medical care in that way wasn't even a consideration."

Nowadays, when Lydia goes to the hospital, Brisky often goes with her, and the two work together to best accommodate Lydia's needs (things like going to a quiet, dark place rather than a regular waiting room). This reduces sensory overload and can also help with her erythromelalgia to some extent, since her symptoms make it harder for her to answer questions or handle being touched. Sometimes, Lydia can just look at her mom, and Brisky will know what to say.

"As an older person, you tend to put your foot down more," Brisky said. Lydia's descriptions of the hurdles autistic people face with health care resemble the predicaments that many face when asking for academic accommodations: they often put the onus on the autistic person to ask for them. In the case of the health-care system, autistic patients and their families must advocate for themselves, and individual doctors might not be empathetic to their needs. We should demand that doctors' offices and hospitals be built with autistic people in mind.

Doctors' misconceptions about autistic people can have severe, life-threatening consequences. The National Council on Disability released a study in 2019 that chronicled the experiences of Paul Corby, who is autistic and has psychiatric disabilities. When his mother, Karen, reported Paul's worsening physical condition, which included "vomiting, chest pains, and a persistent cough," doctors attributed those symptoms to anxiety and failed to evaluate his heart. Paul's mother said he was denied proper treatment because he carried a doll for comfort and failed to name all nineteen medications he took.

Paul Corby's story made national headlines, appearing in the *Washington Post* in 2017, when it was reported that he was ultimately denied a heart transplant because of his "psychiatric issues, autism, the complexity of the process and the unknown and unpredictable effect of steroids on behavior." When I reached out to Karen Corby in July 2020, she told me he had still not received a transplant, though his condition is stable for now.

Autistic people's fate often depends on whether they can find a doctor who will be receptive to them. For Lydia, her emergency department care improved after she contacted Dr. Arvind Venkat, an emergency physician at Allegheny General Hospital. She had read comments of his about the challenges autistic people face in the emergency department. When the two of them spoke, he asked her what the biggest problems were for autistic patients, to which Wayman said sensory overload and communication.

Lydia said that Venkat specifically intervened with the emergency department and spoke with both the medical and the nursing director. As a result, when Lydia must go in for a visit, she has fewer people tending to her at a time.

"So, instead of a doctor, a resident, and three interns asking questions while one nurse is trying to start an IV while talking to another nurse and someone is asking me to sign, things happen one a time," she explained. For instance, if a department is chaotic, an X-ray technician and lab people will come to her rather than leaving her sitting and waiting in the hallways.

"They slow down, speak softer, and ask before they touch me. They ask a question and then wait for an answer," Lydia said. They let Brisky stay and help when needed, but now they speak directly to Lydia, not around her. "They accommodate me but treat me like I am capable."

Since the intervention, Lydia said the difference in her ED care has been like "night and day," and she means this literally, in a way—the hospital dims the lights for her.

Dr. Arvind Venkat said that he received little training in medical school about handling autistic patients. Over time, as he started seeing more and more of them, he noticed that the emergency system was ill-prepared to handle them.

"If you are purposely designing a place for people with ASD, you could not do worse. It is loud. There's bright fluorescent lighting," he said.

To better aid autistic patients, Venkat has collaborated with Joann Migyanka at the Indiana University of Pennsylvania on ways to better educate doctors about autism. He also reached out to the University of Pittsburgh and Indiana University of Pennsylvania to educate emergency medical service providers and doctors.

"We've published a lot on that area about, not just the medical problems that can bring individuals with autism spectrum disorder to the emergency department, but also the special needs of these patients in dealing with what, by definition, is a crisis situation and can be extraordinarily challenging beyond what others may experience in the emergency department," he said. Venkat noted that one of the challenges when it comes to autistic people is that despite the greater recognition of autism in the health-care community, most autistic individuals are still not getting care specialized clinic setting.

"Even when doctors and nurses have some understanding of autism, there's nothing [systemic] in place in these hospitals," Lydia said. "We're not teaching medical students," adding that medical students need a social worker who is a liaison to people with disabilities because "there's zilch in a hospital."

Lydia's ordeal illustrates just how difficult it is for autistic people to get proper medical care but also that there's hope if they can find a doctor who is willing to accommodate them.

Complicating the problem is that America knows little about autistic adults. In May 2020, the Centers for Disease Control and Prevention released its first study on autism in adults, which estimated that 2.2 percent of adults were autistic. But the study was only an estimate and was the result of using the estimates of autistic children and then projecting that onto the adult population while also incorporating data on mortality. There are still not enough data about autistic people's health concerns, which would be beneficial for autistic adults and for parents of young autistic children. For instance, what little research exists about autistic people's life spans shows that autistic people do have health problems that need addressing. The National Institute of Neurological Disorders and Stroke notes that 20 to 30 percent of all autistic individuals will develop epilepsy by adulthood. Given the potential for sudden unexpected death in people with epilepsy, this is an area that requires more in-depth research.

Furthermore, the little research that has been conducted shows that there is a raft of comorbid health conditions that affect autistic people, but much of the research has been conducted outside of the United States.

One study from 2015 funded by Sweden's Karolinska Institute and the city of Stockholm surveyed more than 27,000 autistic adults and compared them with 2.7 million nonautistic adults. The report found that the group of "low-functioning" autistic people had a higher mortality rate than that of the group of "high-functioning" autistic people but that both had a higher mortality rate than the general population. Furthermore, the leading cause of death for the "low-functioning" group was epilepsy; the most common cause of death for autistic individuals without intellectual disabilities was circulatory conditions, like heart disease, followed by suicide.

Of the $364,435,254 spent on autism research in 2016, only 2 percent went toward researching life-span issues. By comparison, that same year, the United States spent 35 percent on autism and biology and 24 percent on risk factors, which included the role of genetics, epigenetics, and environment in autism's development and the interaction of these factors. Combined, these two areas of research account for more than half the money spent on autism, yet they are mostly about what causes its symptoms. That means only a small fraction of the money goes toward helping autistic adults as they age.

Budgets reflect values. Where you choose to spend money indicates what is important to you. Plenty of autistic people worry that a focus on biology and genetics is essentially a focus on eliminating or mitigating autism. To them, all this focus on biology means that politicians and lobbyists still aren't listening to their needs. That, combined with the fact that so little money is spent on life-span issues, proves that the United States is more interested in finding out why people are autistic and less interested in understanding what autistic people need right now. Studying biology is not inherently bad, but this disproportionate amount of funding is an example of how America is talking past autistic people.

When autistic people are consulted, they can offer new insights into a better approach to health care for their community. That was the case in 2012 when the *Journal of General Internal Medicine* published a study in which a majority of the researchers were autistic. They used a community-based participatory research (CBPR) model that surveyed 209 autistic and 228 nonautistic individuals. The study found that autistic adults had lower composite scores for patient-provider communication, general health care, and chronic condition self-efficacy than their nonautistic counterparts. Furthermore, autistic adults were less likely to have received a tetanus vaccine and, if they were female, to have received a Pap smear in the previous three years. The study also found they were twice as likely to have used the ER in the past year. The researchers said it could provide a template for future research.

"Our CBPR process allowed us to adapt instruments to be accessible to this sample of autistic adults, primarily by increasing the clarity and precision of the language used in the instruments," the study said.

But physical health is only part of the equation for autistic people, and it is possible that autistic people's mental-health needs might be more acute than neurotypical people's needs. One 2018 meta-analysis of sixty-six studies in the *Journal of Abnormal Child Psychology* found that people with autism spectrum disorder "are 4-times more likely to experience depression in their lifetime." Suicidal ideation is something I am all too familiar with. I've lived with depression as long as I can remember and have been on numerous antidepressants over the years. When I find myself in the grip of a depressive episode, it is nearly impossible to get out of bed. (The whole myth that depression makes you a good writer never made sense to me; how can you write when you can't even get out from underneath the covers?)

When I feel like I'm facing intractable obstacles—be it having my heart broken, feeling socially isolated, feeling like I let down my family and friends, or just feeling like I want the pain to stop—I have on occasion just wondered whether it was worth it to keep going when pain and despair felt insurmountable.

Thankfully, every time, my other senses have prevailed, mostly out of obligation toward other people. The unfathomable despair I felt after my first year of college was supplanted by pride when my sister, Stephanie, got into Loyola Marymount University, which made me want to see her graduate college. When I thought about suicide as a student at UNC, I remembered my friends at the *Daily Tar Heel.* The times I was nearly suicidal because I had gotten myself into some mess with some woman (most of the time it was my fault and I had put too much pressure on her), I would call the suicide hotline. (If you are based in the United States and feel suicidal, call the National Suicide Prevention Line at 1-800-273-8255. I have called and it helps.)

At the same time, I worry for autistic people who don't have the same support systems I do. Many are kept isolated in institutions and are viewed as unempathetic and unable to feel for others, which only adds to their feelings of loneliness. It is my bonds with other people, my sense of duty, and my ties to those around me that saved me. Thinking of autistic people who died by suicide, I wonder if a community could have saved their lives the way it did mine.

Only in the past six years have I started seeing a therapist and only even more recently have I started talking more openly and frankly with her about my depression and the things that drive it and fuel my self-destructive behavior.

I know I am fortunate because I am alive. Cal Montgomery noted that we don't know what autism in and of itself looks like; we only know how autism informed by trauma presents itself. Whether it is autistic people who don't speak or those with intellectual disabilities who have been subjected to shock therapy or those, like Lydia Wayman, who have lived through nursing homes and doctors questioning their agency, that trauma becomes inextricable from our lived experience. It is sewed into our identity as autistic people. That trauma won't disappear automatically. It will last and linger. But instead of discrediting our experiences, health-care professionals need to honor and respect our trauma in order to help heal all those injuries that the world has inflicted.

6

"Ain't Talkin' 'Bout Love"

RELATIONSHIPS

I arrived in Newark, California, a suburb of San Francisco, on a Sunday afternoon, which is pizza day at the Williamses' home. Chris and Cori Williams's kids, Calvin, Charlotte Ruby, and Cassidy Luna, all autistic just like their parents, greeted me by rattling off different breeds of cats and dogs. I met their two dogs and their twelve-week-old black kitten while they were watching YouTube videos on the living room's large-screen television. Chris joked with me that while some autistic kids didn't speak, some were hyper-verbal, and his daughter Charlotte definitely fell in the latter bucket.

When Chris and Cori met, in 2004, neither of them knew they were autistic. At the time, Cori was living in the United Kingdom and seeing another person, but she was planning on returning to her home state of Utah, where Chris was living at the time, when she created an online dating profile. Chris said there was an instant connection.

"All the ingredients were there for me at the very beginning, just in terms of feeling comfortable, feeling excited, just feeling very happy to be in her presence," he said. For Cori, the first thing that stood out about Chris was his voice.

Because of the time difference, their messages would be the first thing they'd see in the morning. "He would send me an e-mail when I'd be sleeping, and so I would wake up to an e-mail where he had typed out some e-mail about a song, and then I would listen to the song," Cori said. In addition, they would exchange music, like the Verve's "Drugs Don't Work," "You'll Accomp'ny Me" by Bob Seger, and "The Blower's Daughter" by Damien Rice. I could tell that they enjoyed reminiscing about those early euphoric days when you get to know the person you'll eventually love.

They stopped talking for a while, but Cori returned to Utah after her mom's brother died. Around that time, Chris messaged her, and she remembered how much she liked his voice and his taste in music. What really sold Cori on him was that shortly after they began dating, Chris started to help out her sister, who was a new mom and also had a four-year-old. Chris and Cori would send her sister to bed and then put the baby to sleep together. She remembers leaving the room to make the baby a bottle of milk and coming back to find Chris totally enamored. "He was like, 'She's so tiny. She's so beautiful,' and I was like, 'I'm going to marry that boy.'"

They got married in 2007.

In many ways Cori and Chris's romance is a typical boy-meets-girl love story, which is the case for many autistic couples who date in the same ways neurotypical people do. However, because modern dating is dominated by unspoken rules and vague social cues, some autistic people face a unique set of challenges. For many couples—Cori and Chris included—there are some explicit benefits to online dating as opposed to real-world dating. For one, the people using dating sites or apps like Tinder or Bumble are there for an explicit reason, thus removing the ambiguity that might come with flirting.

One survey of seventeen autistic individuals released in *Sexuality and Disability* found that six participants thought online dating was easier and four said there was no difference between meeting in person and online. But while participants said they enjoyed the convenience of online dating, like the ability to fill out their profile information to have it readily available during the dating process, one of the concerns many cited was a lack of safety because of the risk that comes with meeting strangers online, particularly "people presenting false images," which is "troublesome especially for Aspies as [we] tend to be a bit naive and trusting."

Chris came to learn he was autistic when Cassidy Luna was diagnosed. They began searching for as much information as they could about autism, including reading Steve Silberman's book *NeuroTribes* and following autistic voices on Twitter, Chris started to recognize some of the hallmark characteristics of autism in himself. When Chris heard an episode of the comedy podcast *Aut-tastic,* he studied himself, spoke to his wife, and finally in 2016 went for a "battery of tests and interviews."

"It was a powerful moment. I think I kind of felt seen in ways I really appreciated," he said.

Cori has strawberry-blond hair, which she pulled back into a ponytail, and was wearing blue flip-flops during our interview. Cori was the last in the family to realize she was autistic, and though she is self-diagnosed, she also learned that she might be on the spectrum through her daughter Cassidy's process. While researching autism on behalf of her daughter, Cori saw a video about how autistic females mirror (which is to say follow) others' traits, which made her think, "Holy crap, I might also be as well." She said her upbringing in Utah was abnormal, but she never considered the reason behind it. "I just struggled in school socially, for sure; it was awkward, but I attributed it to being poor and non-Mormon in a rich Mormon area," she said.

Despite families likes the Williamses', years of misconceptions have made autistic people seem incapable of being good partners. Much of this is rooted in the idea that autistic people are incapable of empathy, which is simply not true. In 2007, *USA Today* quoted Katherine Tsatsanis of the Yale Developmental Disabilities Clinic as saying, "Almost by definition, an Asperger's person would not form an intimate relationship, get married and have children," since "they don't form connections. The desire, the drive and the social knowledge is lacking."

When autistic people do try dating, it is often treated as a curiosity for people to gawk at, as was the case with Netflix's documentary series *Love on the Spectrum,* which chronicled autistic people in various settings trying to date and find love. There are countless stories about neurotypical people asking their autistic classmates to prom, which often makes dating autistic people seem like an act of charity performed by popular cheerleaders or jocks helping a poor, unloved autistic person.

But this narrative that autistic people are incapable of relationships or get dates only out of pity is simply false. Plenty of autistic people do have partnerships and often families that thrive specifically because they understand each other's ways of functioning in the world. Both Chris and Cori say that there is a benefit to their both being autistic and synchronizing on the same level, though they are neurodivergent and autistic to different degrees. Cori told me that she thinks that people in her past relationships were likely autistic.

"I do think that you're attracted to the quirkiness when you have it by yourself," she said. "Autistic people sometimes have this need for sameness, and I think those are things that did attract us to each other."

Seeing the way Chris and Cori work together as a parenting unit and a couple contradicts the conception that autistic people cannot form relationships. The idea that autistic people, particularly autistic men, are incapable of finding love is at best fodder for countless television shows and movies, and at worst it feeds into a toxic stereotype about autistic men who will eventually turn on women. Furthermore, stereotypes about autistic people's lack of capacity for empathy and caring have created pernicious tropes about their inability to understand consent and sex and spreads the idea that they are unable to adequately parent children.

"How Do I Know When It's Love?"

The idea that autistic men are somehow awkward, cold, unloving, or unable to interact with women goes back to before the public understood autism as a spectrum. In 1984, therapist Jean Hollands wrote a book called *Silicon Syndrome: How to Survive a High-Tech Relationship,* which focused on how women could navigate "high-tech relationships" with "sci-tech" men who "tried to address problems in intimate relationships by seeking data." The book's introduction lamented husbands who appeared more interested in how computers worked than in their wives. While the book was released before the understanding of autism was expanded to include what would later become known as Asperger's syndrome, Steve Silberman wrote that "Hollands could have swapped the term Asperger's Syndrome for Silicon Syndrome and barely changed another word in the text."

Since then, television shows like the Netflix series *Atypical* revolve around autistic people—mainly autistic men—struggling to navigate dating. The problem with these portrayals is that they often give autistic men a pass for treating women terribly. In one of the early episodes of *Atypical,* the autistic protagonist Sam shuts his girlfriend Paige in his bedroom closet after she holds his pet turtle. Later on, Sam breaks into the house of his therapist, on whom he has a crush. While both scenes are meant to elicit laughter, they perpetuate the stereotype that autistic men don't know how to treat women with respect.

What's worse, these lighthearted portrayals let autistic men off the hook for their bad behavior. Certainly in my own life, I have not always treated women with respect, but I try to hold myself accountable nonetheless. Going back to when I was in fifth grade, I would send long-winded letters to girls I liked. Later, in high school, college, and even into adulthood, if I liked a woman, I would try to hang around her and shower her with gifts or bombard her with texts in hopes she would eventually like me back.

The thing I most regret was how I treated one of my first roommates in DC when I moved here full-time in 2014. We initially hit it off as friends, but I wanted more. Even after she told me I was becoming overbearing, I continued to pursue her, and ultimately, I wound up driving her even farther away. I also regret that this behavior continued for years with other women, to varying degrees. It would be easy to simply chalk this up to me being autistic, to not understanding social cues or not being able to read women's minds. But that would be using autism as an excuse; it would imply that autistic people are incapable of intimate relationships with neurotypical people because we will always blow them up with our own behavior.

The truth is the mistakes I've made with women are not directly caused by my autism. As I started to address my relationships in therapy, I connected this behavior to my childhood. Growing up, I was incredibly lonely. In the first ten years of my life, we moved from Chicago to Florida to Wisconsin to Northern California to Texas and then finally settled in Southern California. I was also bullied constantly when I was growing up. I still remember when my classmates in fourth grade coaxed me into moving the front of my body up and down a tree; I didn't realize it was something called "humping" and that it was sexual until much later. When I was a freshman in high school, in an effort to fit in with kids, I frequently picked on other kids, harassing, hazing, and humiliating them. In one of these cases, I was the kid being hurt, and in the other, I was doing the hurting, but at its core, I was desperate for any kind of love and friendship.

I should add that my home life was great and my parents loving. But when they divorced, it wasn't easy for me. I frequently felt torn between the two, like I was forced to pick sides. Eventually, my mom began dating Bob, whom she was with for years, and even though they never married or lived together, I thought of him as my stepdad. Still, it hurt that after we had great time going to museums or a concert or just spending a weekend together, he'd leave. That feeling of being an outcast at school while feeling torn at home made me desperate for love in any form. Like many straight teenage guys, I wanted a girlfriend but didn't really know how to talk to women.

At the same time, I had heard that the likelihood of autistic people getting married was low, which depressed me because I wanted to be loved. What if my autism meant I would never find anyone who would accept me for me? This makes my reporting about and furthering my own understanding about autism all the more ironic. Reporting this book and integrating myself into the larger community shattered my own assumptions of what autistic people were capable of doing. I've met people like Chris and Cori, who both have lower support needs but have kids who require more support. I've met couples with mixed levels of support needs who dutifully support each other, and I've met couples where one person is autistic and the other is neurotypical. This has allowed me to see what potential dating and marriage scenarios might be the best fit for me. As of right now, I am still undecided about family life. Like many kids of divorced parents, I take the decision to have children very seriously and don't want to put a child into a scenario where they might be in a broken household. I also want to make sure I am financially stable, mentally mature enough, and willing to devote my entire life to any offspring. I want to be present physically and emotionally for any child I have. That being said, I am keen to at least find a partner to go through life with.

I often used to find myself looking for love in places where I shouldn't. Often, if someone showed me any measure of kindness, I would mistake it for romantic affection and go all in. Or I would ask a woman out for coffee or drinks thinking it was a date without realizing that she wasn't on the same page as me. I know now that this was not the women's fault; it was my mine for not being clear about my intentions.

Once I did wind up going out with a woman, I found myself overwhelmed by the expectations of dating, the proper etiquette, or what to say to a woman, outside the normal rules of "just be kind" and "don't be a jerk." While I am trying to enjoy the date, I am also trying to pay attention to what my date says and remember things she likes. Generally, I'm trying to see if she is a compatible partner, but I also want to make sure that I am paying attention to her interests and what matters to her. And this is to say nothing about the difficulty that comes with disclosing that I am autistic. Before I started writing about autism for a wide audience, on every date I went on, I knew eventually I'd need to find a way to disclose. Then, when I started writing about it, any woman with even a slight amount of skill at Googling could find out that I was autistic.

In some ways that liberated me because I no longer had to explain, and it filtered out women for whom that was a deal-breaker. But it also made me wonder just how much of myself I could be. Everyone frets about the "just be yourself" paradox of first dates, where you have to decide which parts you want to save until the second or third dates. And for autistic people, particularly those who defy expectations, the question of how much of yourself you should reveal is even more fraught.

Because I can speak and don't "look autistic" (whatever the hell that means), people often think my autism is negligible and doesn't define me. But it is still part of me. As a result, I worry just how autistic I can be on dates. For example, if I get excited talking about a subject, I worry I am dumping all the information I know on my partner (and if it's a topic on which my date is an expert, I absolutely do not want to condescend to her by overexplaining). If we are seated across from each other, I worry she will take offense at me not making eye contact. Online dating during the coronavirus pandemic had definitely sucked, but one good thing about it was that it made eye contact almost a nonissue and I could fidget with my keys or a pen as a stim.

In short, masking and blending my autism makes dating difficult. I am used to hearing people say, "You don't seem autistic." When I was younger, I took it as a compliment. After being teased and singled out for being strange, hearing I blended in was the highest form of praise because it meant I belonged. But now, I resent that description because it means that I don't fit people's stereotype of autism. Trying not to "seem autistic" is a reaction to the fear of being "caught," and that blending means you never know when to show your partner the real you.

Even though Chris Williams said it was likely that his previous romantic partners were neurodivergent in some way, he still struggled with how much of himself to reveal to them.

"It felt like I had to mask more, even though I didn't have that word in other social settings," he told me. But when he was with Cori, he felt he could be himself since he received love and affection from her. "I felt like our authenticity with each other was reciprocated and was appreciated, and I certainly think that helped us just kind of settle into each other very quickly," he said of his relationship with Cori.

Cori said that before autistic people realize they are masking, it is hard to feel accepted. She said when she was with her previous partner in England, she had to be a completely different person and reinvent herself, and as a result, she wore that mask for a long time when she started seeing Chris.

"It's something that I think about from time to time," she said, speaking to Chris. "Like, did you fall for my mask in some ways?" Chris responded, "Yeah. It'd be interesting to ponder that."

Ultimately, my life and my relationships with people improved when I got serious about going to therapy and being transparent with my therapist. That led me to take a critical look at the way I had treated women in the past, how I approached both friendships and romantic partners. This in turn led me to listen and make amends to the friends and other women in my life. I stopped trying to force women into liking me. I stopped trying to smother them in the hopes they would reciprocate my affection. I became more okay with rejection when things didn't work with someone I genuinely liked. Essentially, I worked on becoming someone *I* would want to call a friend or partner. Contrary to popular belief, autistic people are not unempathetic, and multiple studies have debunked that myth.

Once autistic people discover themselves, become at ease with their condition, they can find fulfilling and successful relationships, like Anlor Davin and Greg Yates, who are both autistic and live in the Bay Area, did.

Davin—who, when we met in a café in San Francisco, was wearing a black blazer with a cap evocative of her upbringing on the west coast of France—spent years trying to figure out what made her different from others. She said that her autism was overlooked because of her gender and that when she was finally diagnosed, it drove her to tears. "It was not tears of unhappiness, it was like, 'Yes. I understand.' And then the pieces of the puzzle started to get together," she said.

Yates met Davin at a picnic held by the Autism Asperger Spectrum Coalition for Employment and Networking Development (AASCEND), a nonprofit that focuses on supporting autistic people in the

Bay Area that Yates cofounded. Incidentally, the first thing Davin did after she was diagnosed was go to an AASCEND meeting, and she said it was "a real relief" to be around other autistic people.

"I mean, we understand one another even if we're different, all of us. We still have some similarities including main things, like the sensory stuff," Davin said. Davin said it was an anxious time for her and she was also grateful she was not diagnosed in her native France, a country whose treatment of autistic people has been denounced by the United Nations and where parents are often pressured to send their children to institutions.

But just because two people are autistic does not mean that they will gel, the same way that dating within one's own race, class, or religion doesn't guarantee success. Yates says one thing that helps with their relationship is the fact that they are very different people.

"Our autisms are not the same but there is some overlap," he said. "And so one of the things that keeps us together is the fact that we're both autistic and we recognize how much of the challenge of autism can be hidden."

Before Yates met Davin, he was married to a neurotypical woman with whom he had two children. He'd always wanted to have a relationship and figured that since he was smart (he has a degree in biophysics from Berkeley and was briefly in a PhD program at MIT), he could synthesize social behavior with his intellect. "Like a little hamster on a wheel, I worked hard to keep that illusion up," he said of this relationship. Once he began to meditate, however, something shifted, and Yates felt he no longer had to hide his autism from his first wife.

Meditation plays a big role in Davin and Yates's relationship. Davin practices a style called Soto Zen, and the two lead a meditation group called Autsit. Davin agrees with Yates that meditation plays a major role in their relationship. She also says that one benefit is they understand how each other's brains work. "Well, for me, there's a respect of the other person in terms of their sensory stuff or their giving facts all the time," she told me.

Throughout the discussion, Davin and Yates occasionally huddled and discussed things between themselves and made sure they were on the same page. Plenty of times, they finished each other's sentences or helped clarify a detail, like the exact time they met. At one point, Yates mentioned that they recognized that their challenges could be hidden and tried to be sensitive to the other person's needs, and Davin told me that at times Yates slammed doors, which was hard for her sensory processing.

"So we both have an intuitive understanding of the weight of what we are all carrying in our lives because of the autism," he said. All this is to say, they are just like any other married couple. They recognize they are both autistic, which gives them empathy, but they also recognize how their individual challenges differ and try to navigate them.

"Human Touch"

The belief that autistic people are incapable of having relationships means that autistic people are often not taught about consent and bodily autonomy. Humorous portrayals of autistic men navigating romantic relationships as in *Atypical* overlook the severe consequences that autistic people can face in the real world. The *Sun,* a newsletter for Autism Delaware, explained some of the many reasons for misunderstanding around autism and sexuality in its Summer 2017 issue.

"Fears and stereotypes often impede individuals with disabilities," the newsletter said. "Stereotypes include the belief that people with ASD are childlike, nonsexual, over-sexual, unable to understand, unable to give consent, uninterested in sexual relationships, unable to develop or maintain a sexual or romantic relationship, or not able to get married or have children."

These misperceptions make it all the more difficult for people on the spectrum. If autistic people are all childlike or nonsexual, then there is no reason to teach them about consent, because there is an assumption that they will never experience sex. If they are over-sexual, then teaching them about sex may ignite their overactive libidos. If they are unable to give consent, then teaching them harms their safety. All of these misperceptions are rooted in the idea that autistic people need protection from neurotypical people. Caretakers must either protect autistic people from an overly sexualized world that can prey on them or defend the world from a hypersexual autistic person. As a result, when men like Sam in *Atypical* break the law or violate a woman's privacy, society often gives them a pass.

Amy Gravino, the autistic relationship coach at Rutgers University, said of Sam's indiscretions on the show: "[For] this young man, there are no consequences. But that's not reality. If you commit a crime, a sexual crime, you will go to jail."

Autistic people have been fighting the stereotype that they don't understand consent for years. Before the 1970s, autistic people were mostly thought incapable of having relationships. And to this day, little (if any) attention is paid to autistic people's sexuality. Research confirms that parents underestimate their autistic kids' sexual experience. A survey of forty-three parents and their autistic sons in Belgium and the Netherlands in the *Journal of Autism and Developmental Disorders* found that half of the parents surveyed said they did not know that their children had masturbated or experienced an orgasm. Conversely, very few parents overestimated their neurotypical sons' experiences. Gravino emphasized that one problem with teaching consent is that autistic people are often infantilized, so they typically don't receive proper sex education.

"So, when you're always seeing someone as a child, you're thinking that they're innocent; you're not seeing them as a fully realized adult with the same desires and needs as other neurotypical adults," she said. "And so parents will say, 'Well, no, if I tell my child about sex, they're going to be thinking about sex and want to have sex.' And it's like, 'I can guarantee you they are already thinking about it.' They're not going to suddenly put ideas in their head."

Much of the fear around teaching autistic people about sex and consent assumes that they don't have the same sexual desires as nonautistic people or are completely asexual. It is also due to the fear that they could either get hurt or make bad decisions. But while some autistic people are indeed asexual (just like some neurotypical people are asexual), the decision to have a sex life should not be made for them any more than it should be made for nonautistic people. But much like the right to live independently, autistic people's right to have control of their own sexuality means they have the right to make poor decisions, get their hearts broken, and fail as much as neurotypical people. (To be clear, when I say "bad decisions," I do not mean sexual assault. If an autistic person inflicts harm on someone, I believe they should face the consequences of their actions as any person should. "I'm autistic" is not an excuse.) Autistic people have the right to make mistakes as much as their neurotypical counterparts.

That said, some believe that there needs to be a sex education course designed with autistic people specifically in mind. "Because we tend to learn differently, the sexual education curriculum programs that are geared towards neurotypical individuals may not really work for us in terms of being able to effectively learn things like consent and boundaries and so forth," Lindsey Nebeker, a development specialist at the Autism Society of America who is autistic herself, told me. "And for me, I know, learning about boundaries and consent was something that I didn't pick up on."

Nebeker and her husband, Dave Hamrick, are perhaps one of the most prominent autistic couples in this country. Since the two met in

2005, they have been featured on *Good Morning America* and were the subjects of a documentary called *Autism in Love.* When I visited Nebeker and Hamrick in 2016 for an article I was writing for the *Washington Post* on sex and consent, they were in their first year of marriage but had been living together after two years of dating.

One of the problems when it comes to autism and consent is that for most people, autistic or not, their understanding of sex—and therefore consent—is shaped by their surroundings and environment. In Nebeker's case, her parents didn't necessarily relay in-depth information about sex to her (growing up in a devout Pentecostal family where I had only one "talk" with my dad while attending strict Christian high schools that taught abstinence over biology, I can relate).

"We didn't have a lot of those conversations and that's why, as I was growing up, I would have issues with knowing what consent actually meant," Nebeker told me.

Like a disproportionally high number of autistic people, Nebeker is a survivor of sexual abuse, which took place at the hands of a high-school teacher while she was in boarding school.

"That was probably the greatest example of my life of being taken advantage of," she told me. Nebeker said it was a confusing relationship and she still questions if her teacher knew what he was doing. "I think even throughout that and being hurt by what happened, there's a little part of me that still loves him." Many survivors of abuse can feel conflicted emotions about their abusers and Nebeker is no different.

Nebeker said one of the reasons she might have been a vulnerable target is that, like many autistic people, she had difficulty finding acceptance as a child, so she craved validation and affirmation. "I think that gets forgotten with our autistic population," she told me. "I think one of the reasons I had fallen into being taken advantage of is because I didn't feel I was worthy of being loved."

As a result, when Nebeker is intimate with Hamrick, they have to navigate her trauma with caution. "He always, every time before we get sexually intimate, he always asks me first," Nebeker said. Hamrick echoed this. "It's like, when I go into her intimate space, it's nice to have the consent to do it," Hamrick told me. I find this to be one of the most romantic things any partner could do; love is simply wanting what is in another person's best interest. Nebeker and Hamrick show how respectful autistic people can be when it comes to sex, a contradiction to the idea that autistic people don't understand sensitive subjects like consent.

"Could You Be Loved"

The media's single-minded focus on autistic men has often excluded LGBTQ people, who make up a sizable portion of the autistic community.

One study from *Autism Research* from 2018 surveyed 309 autistic individuals and 310 typically developing individuals and found that while 30.9 percent of the typically developing group reported being nonheterosexual, 69.7 percent of the autistic group reported being nonheterosexual, with higher rates of bisexuality, homosexuality, and asexuality. The monolithic portrayal of autistic people's sexuality prevents LGBTQ autistic people from navigating their dual identities.

Gravino said this exclusion erases queer people from discussions of autism, dating, and sex.

"Imagine you're on the spectrum and you want to be accepted as being a sexual being and an adult, and then you're LGBTQ on top of that. That's a whole other kind of barrier and challenge that a lot of folks end up facing," she said.

Indeed, plenty of autistic people interviewed in this book identify as LGBTQ. Jessica Benham, the autistic state legislator, is bisexual and married to a man. Julia Bascom, the executive director at the Autistic Self Advocacy Network, is a lesbian. John Marble, the former Obama administration official now dedicated to helping autistic people find work, is openly gay and worked for the national Stonewall Democrats organization, which promotes LGBTQ rights.

Marble said that coming out as gay informed his own experience of understanding his autism. "I'm glad I had that coming-out experience as gay first because it gave me a reference point," he told me. It taught him that it is okay that he is different and that, much like being gay, he would find people like himself. But at the same time, "You don't have to get super-enthusiastic and go to like every autism march. It was like a tap on the hand," he said.

M. Remi Yergeau, the autistic University of Michigan professor, said that often, the world questions when autistic people identify as nonheterosexual or anything other than cisgender, and they chalk it up to autism. "So, if you're trans, if you're queer, if you're anything under the LGBTQ umbrella, clinically it's seen as, "Well, that's because you're autistic, and you just don't know better. You don't have a social compass, because you're autistic. So why would you know that you want XYZ?" Yergeau said this questioning is backdoor homophobia and transphobia.

Incidentally, efforts to force autistic people toward neurotypical standards and LGBTQ people toward heteronormative standards share the same DNA. Ole Ivar Lovaas, the researcher who pioneered work on applied behavior analysis and who ran the UCLA clinic where autistic children were slapped, also assisted with UCLA's Feminine Boy Project and coauthored four papers on atypical sexuality.

In the first study, released in 1974, Lovaas and George Rekers conducted their work on a five-year-old boy they called Kraig whose real name was Kirk Murphy. Early in his life, Kirk displayed a tendency to dress in girls' clothes and play with dolls. When Murphy's mother saw a television ad for a program promising to "fix feminine boys," she took Kirk.

During the treatment, Kirk was given the option of playing at one of two tables. One of the tables held traditionally masculine objects, like a rubber knife, cowboy and Indian toy sets, handcuffs, and a dart gun; the other table held girls' clothing and grooming toys. When Kirk played with feminine toys, Kirk's mother was instructed not to speak to him, but when he played with the masculine toys, she would praise him. At home, when he displayed masculine behavior, he was awarded blue poker chips that could be used to watch TV or get candy; when he exhibited feminine behavior, he was given red poker chips that meant he would be spanked by his father. Kirk's brother Mark later told CNN that seeing his brother's abuse was so unbearable that Mark would occasionally hide the red chips to prevent his father from giving Kirk spankings. The effects on Kirk were so devastating that in 2003, he killed himself, which his brother and sister blamed on the traumatic experiences from his treatment.

All the while, Rekers and Lovaas hailed "Kraig's" transformation from a boy who would wear girls' dresses and even improvise by using towels when they weren't available to a boy who wore sneakers and blue jeans. Rekers later cofounded the Family Research Council, the socially conservative group that opposed same-sex marriage. In 2010, he was photographed at an airport with a male escort from Rentboy .com.

Kirk's story and his subsequent desire to mask his homosexuality through traditionally masculine acts such as joining the air force mirror the stories of autistic people who felt they had to blend in and be neurotypical or jump through the series of hoops to please other people. In both cases, they were desperately trying to please others and their parents were doing what they thought was best. But when the parents of LGBTQ or autistic children prioritized their own fears and anxieties over their children's happiness, they wound up losing their children.

Much of the language of those who discriminate against LGBTQ people matches the rhetoric of those who discriminate against autism. Joseph Nicolosi, one of the most notable practitioners of conversion therapy, told CNN in 2011 that "homosexuality is an adaptation to an emotional breach with the parents," usually with the parent of the same sex. Listening to Nicolosi, one cannot help but hear parallels to Leo Kanner, who in 1960 said that parents of autistic children just happened "to defrost enough to produce a child."

But in both cases, parents, clinicians, and researchers were trying to make these children worthy of love by removing the very things that defined them. When parents make autistic kids not flap anymore or boys wear jeans instead of dresses, they replace the child that exists with the one they wished existed. They sentence them to the misery they allege they want their children to avoid.

The fate of autistic and LGTBQ people are intertwined (to say nothing of autistic LGBTQ people), as both groups' right to happiness on their terms are regularly suppressed. They are told they cannot be loved unless they abide by the terms the world has prescribed. But they are worthy of love as is. It is when they come to accept themselves that they recognize this truth. In the same way, I can say that when I finally shed my own disgusting homophobia—which was a means to protect myself from other people's torments and was fostered in chat rooms on social media sites like Myspace—and befriended LGBTQ people, I began to hate the autistic parts of myself less. None of us are failed versions of normal. We can love and be loved as is.

"Come Mothers and Fathers Throughout the Land"

In the same way that people worry about autistic people's capacity for consent in sex, there is also this belief that autistic people are incapable of raising children or being good parents. The idea that autistic people could even be parents was dismissed for years; in 1988, researcher Edward Ritvo published a paper called "Eleven Possibly Autistic Parents." He told *Spectrum* in 2017 that if he hadn't included *Possibly* in the title, the paper would likely have been rejected. "Nobody believed it. They didn't believe the parents had it, that autistic people could grow up and marry and have children."

Sadly, the idea that autistic people can't be parents persists today. In 2016, the writer Judith Newman released the book *To Siri with Love,* an extension of a 2014 *New York Times* article about how the iPhone personal assistant Siri helped her autistic son Gus navigate a neurotypical world. But in the book, Newman wrote, "I do not want Gus to have children," adding, "if I had to decide based on the clueless boy I know today, it would be easy: Gus should not be a parent." Newman said this not just because Gus was "still shaky" about where babies come from "but because of the solipsism that is so much at the heart of autism makes him unable to understand that someone's needs and desires could ever be separate from his own."

While the book received rave reviews and was named a 2017 Notable Book by the *New York Times,* multiple autistic people criticized it for this very argument. The way Newman talks about reproduction is horrific because she considers it her decision to make rather than her son's. This infantilizes Gus and assumes he cannot comprehend a world outside of himself, denying the reality that he could ever care about anyone other than himself. By writing that Gus could never be a "real father," Newman assumes that his disability automatically prohibits him from being a parent. Autistic writer and parent Kaelan Rhywiol wrote in a review for *Bustle* that she'd vomited twice while reading the book, adding, "I cried myself to sleep after finishing it. To know, without any hedging, what people like that think of people like me—it almost broke me."

In 2014, the Iowa Supreme Court ruled that parents with guardianship of autistic children could not arrange vasectomies for them without court approval. The case was raised after a mother in West Des Moines arranged one for her twenty-one-year-old son, Stuart, after she suspected he was engaged in a sexual relationship with a colleague. Stuart and his coworker both denied the relationship. In the end, although the Iowa Supreme Court ruled against the mother, she defended herself by saying, "When I went through guardianship originally, I was never told what I could and couldn't do," adding that her son had the mind of a ten- or twelve-year-old. The fact that this mother believed this decision was her prerogative is indicative of how autistic people are viewed and how their capacity to have real relationships is constantly questioned.

When I am with the Williamses in their home with their kids, there is no doubt in my mind about whether autistic people can or should be parents. To the contrary, Chris and Cori being aware of their autism makes them better parents to their three children; it makes them especially attuned to their kids' needs and also willing to advocate for them when others assume they are not entitled to the same things neurotypical kids are.

Throughout the interview with Chris and Cori, I watched their kids go up to them. At certain points, their son, Calvin, came by asking for help with some gadget. When Cassidy Luna went up to Chris in her pajamas looking for Cori, Chris told her, "Hey, sweetie, Mom is right there," which reassured her because Cassidy is incredibly close with her mom. Still, Cori emphasized that Cassidy's need for more support doesn't diminish her intelligence, competence, or capacity to live the life she pleases.

"The one thing that I do find myself repeating is the assumption of intelligence and competences necessary in raising her," Cori told me. "We don't have any idea what she's capable of yet. She's not going to say very much yet, she's not going to say very much now. But she's got a lot to say and she's going to say it one day. It might not be bubbly but she's going to say it."

Occasionally, Calvin asked for Chris's help with something or one or the other parent left our interview to solve something for the kids. In other words, it was just a normal suburban family with three autistic kids and two autistic parents.

"You've Got a Friend"

Growing up, I had few friends besides JohnPaul, my guitar teacher. His family became friends with my family, and we are still incredibly close. Other than that, save for a handful of people from high school, like my truly close friends Tim and Geoff, or the occasional garage band mates, like my friend Jay, and other Boy Scouts, I felt isolated. But even though my circle was small, it was incredibly valuable.

The first time I developed a tight-knit circle of friends was when I moved to Washington, DC, during my White House internship. This was largely thanks to one of my roommates, Greg, an Irish Catholic guy from Buffalo who was a few years older than me. He would convince me to join him at happy hours after work or parties at our coworkers' homes on Saturdays and then brunch on Sunday. Greg didn't inhale politics the same way I did back then, so we had an unspoken deal that I would give him tips about work and in exchange he would teach me how to enjoy myself socially and talk to women.

Greg and my other roommate, Travis, introduced me to their circles, and eventually their friends became my friends. Those friends loved that I had an encyclopedic knowledge of American politics; all the things that back home in California made me a geek or an oddball were well received in Washington. My friend Raj actually wanted to check out embassies on a Saturday, and my friend Gurwin and I would spend Sundays at a Barnes and Noble reading, ending the night on the steps of the Capitol, where he would teach me economic terms like *short-selling* and *bull and bear markets*. To this day I am still close with many of them.

Still, throughout my life, I often shied away from disclosing that I was autistic to my friends. The first time I did so wasn't until college, when I joined my campus newspaper, the *Daily Tar Heel*. Once again, it was a place where I learned what it meant to have healthy friendships based on a mutual connection; in this case, a love for journalism. Furthermore, most of my colleagues at the paper were women, so it taught me how to have healthy and fulfilling friendships with women, and I ditched the odious idea that solely being friends with women was a death sentence.

Most Thursdays, after we'd put out the paper for the end of the week, we'd wind up going to Linda's, a classic college dive bar with greasy food called Drunchies, a clear nod to the fact students would consume them with some level of inebriation. One night after one of our underage friends was thrown out for sipping beer, we went back to the *DTH* office with booze from the liquor store. It was in that context that I mentioned to my friends that I was on the spectrum and that's why I didn't drive. One of my friends responded, "Me too."

This acquaintance and I had classes together, but I knew him only tangentially. But as soon as he disclosed, I went over and high-fived him, and suddenly opening up became less scary. I realized that my friends would not ostracize me. I would always be welcome at Linda's. Furthermore, these were friends I could trust, and once they knew I was on the spectrum, I no longer had to worry about masking or pretending to act more neurotypical.

The good news is that there are indications that culture is changing and people understand that autistic people can marry, lead fulfilling lives, and have children. In 2019, comedian Amy Schumer revealed that her husband, Chris Fischer, was autistic in her comedy special *Growing*. What was heartening was Schumer did not say that his autism was an impediment to her loving him.

"Once he was diagnosed, it dawned on me how funny it was because all the characteristics that make it clear he's on the spectrum are all of the reasons I fell madly in love with him. That's the truth," Schumer said. She went on to say that he "keeps it so real" because he does not care about social norms, noting, "If I say to him like, 'Does this look like shit?' he'll go, 'Yeah, you have a lot of other clothes. Why don't you wear those?' I'm like 'Okay.'"

Watching Schumer's special, I could not help but nod along and laugh because I recognized so many things in myself, whether it is the idea that autistic people can't lie (here I must protest; we can lie, we just suck at it) or the fact that we don't always have the right reaction when someone we love is in pain but it is not for a lack of empathy or caring.

None of this is to say that being with an autistic person is easy, any more than having a relationship with a neurotypical person is easy. Davin and Yates disagreed with each other during our interview. Nebeker and Hamrick also discussed their points of tension. Chris Williams noted that there are clashes between himself and Cori.

But autistic people desire intimacy and love all the same. We don't want people to love us out of charity or pity. Pity is born out of sorrow or obligation. We want to be loved as equals and for who we are. We want that knowledge that we are loved.

7

"Not Sure if You're a Boy or a Girl"

GENDER

It's the Fourth of July 2019, and I am sitting with Eryn Star (preferred pronouns are *they* and *their*) at a Starbucks in Northeast Washington as they talk about how other people reacted to their being autistic.

Star, who identifies as nonbinary, is in DC to intern at the National Council on Independent Living (NCIL), a disability advocacy organization. They are wearing a gender-neutral T-shirt and glasses. I wanted to meet with Star after I read the piece they had written about how when they were in high school, their choir director verbally abused them. Star is younger than I am and is part of a generation that is much more vocal about discussing the trauma that autistic people face daily and more willing to discuss how autism intersects with other parts of their identity like gender.

The teacher often would say Star was "slower to get things" and berate them for being a few minutes late to class because they were seeing the school psychologist. The teacher knew they were disabled and yet continued to abuse them.

"I think that my experience with that teacher, among other things, may have complicated my relationship with my gender identity," Star told me. "I had other cis-women in my life trying to regulate how I dressed, how I presented my body. And because of that, I do feel disconnected to what womanhood is."

The trauma from their high-school music teacher remained with Star; they didn't listen to choir music or sing for many years. "I've never really been able to let go," they said.

Eventually, Star started taking lessons with a private voice teacher, though they were so tense that their knees shook every time they performed. But interestingly, when Star played the roles of both a man and a woman in a public performance, their knees didn't shake when they sang the male roles.

"My brain was like, 'Maybe you and your gender need to have a chat,'" they said. For Star, their identities as an autistic person and a queer person are inextricably tied. And while Star may not be who most people picture when they imagine an autistic person, their story is not an isolated one.

For years, autism has been viewed as something that exclusively affects cisgender boys and men. To this day, women and girls—not to mention trans and nonbinary people—are overwhelmingly underrepresented in studies about autism, leaving them underdiagnosed and ignored. But overlooking their stories often has horrific consequences that make their lives that much more difficult.

So, for all my difficulties navigating the world as an autistic male, I still get off pretty easy compared to women, transgender people, and gender-nonconforming people. Once it is out that I am on the spectrum, my idiosyncrasies are more or less understood. I'm treated with much more social grace than if I did not inhabit a male body. Autistic women and queer people, however, either face much harsher scrutiny and have their autism questioned or go completely ignored.

"Undercover"

Women have long been involved in autism history. The first major memoir from an autistic adult came in 1986 when animal scientist Temple Grandin published her book *Emergence.* While Grandin's book was hailed, she herself was touted as a source of hope and inspiration for parents of supposedly hopeless autistic children. Grandin's book made her one of the most prominent openly autistic people and, for many years, perhaps the most famous autistic woman in the world.

Still, conventional wisdom is that autism affects cisgender men more than it does other genders. And while it is true that there is a higher occurrence of autism in men—the Centers for Disease Control and Prevention reported that 26.6 out of every 1,000 boys are autistic, compared to just 6.6 per every 1,000 girls—it likely stems from a gender gap in diagnoses, not from any biological predisposition.

Like so many other problems affecting autistic people today, the roots of this disparity go back to early autism studies. Leo Kanner's research featured just three girls out of eleven subjects, though back then, that gender imbalance in a scientific study wasn't as unusual as it would be today. In 1944, Hans Asperger, whose work in Vienna would lay the groundwork for the modern understanding of autism on a spectrum, wrote that "the autistic personality is an extreme variant of male intelligence" and none of the children he reported on in his initial study were girls.

Extreme male brain theory, which takes Asperger's ideas a step further, was proposed by British psychologist Simon Baron-Cohen in 2002. It postulates that male brains are better at "systemizing," or spotting patterns, a trait associated with autism, while female brains are better at "empathizing," which autistic people are incorrectly deemed incapable of.

"Using these definitions, autism can be considered as an extreme of the normal male profile," Baron-Cohen wrote in 2002. Baron-Cohen also argued that exposure to testosterone in the womb might lead to autism.

Many autistic women say these tropes and theories cause harm because they imply that women can't be autistic. "Lots of other women I know and myself are living proof that we're definitely not extreme males," one autistic woman said in a 2017 *Autism* study.

M. Remi Yergeau, an autistic professor at the University of Michigan, is critical of extreme male brain theory. They noted how much of the diagnostic criteria surrounding autism is gendered and how this plays a role in furthering the diagnosis gap. For instance, having "special interests" is a hallmark characteristic of autism, but the common examples given of these special interests are cars and trains—things typically associated with boys. They pointed out that you rarely see ponies used as an example of a special interest.

Some researchers have noted that this might be because girls' special interests tend to be more age-appropriate and less eccentric (think boy bands). So, while an adolescent boy's obsession with trains may signal to his parents that he is on the spectrum, a teenage girl's love of One Direction likely won't spark that same awareness. This disparity is illustrated perfectly in a scene from the graphic novel *Camouflage* in which an autistic girl notes how her special interests were overlooked: "I was told 'all girls like ponies, that's not a special interest' which wasn't true. I knew far more about ponies than anyone else in the riding school."

And gender expectations and perceptions might play a role beyond just special interests. One 2018 study from the *Journal of Clinical Child and Adolescent Psychology* found that girls with ASD had greater social communication problems than boys with ASD. Another study in the *Journal of the American Academy of Child and Adolescent Psychiatry* from 2012 found that girls who met the diagnostic criteria for ASD had "low cognitive ability and/or additional behavior problems" compared to boys with ASD, which suggested that girls were more likely to get missed in the diagnostic process and "may require additional problems to push them over the diagnostic threshold." In short, girls have a higher bar to clear to get an ASD diagnosis.

But autistic girls can face a double bind when it comes to diagnosis. Some autistic behaviors fall under what our society sees as "female behavior" and are thus overlooked as an indicator of being on the spectrum.

"There's certain times when you're quiet because you don't know what to say, [but] then that's okay because that's ladylike," said Morénike Giwa Onaiwu, who is an autistic woman. "If you're super-talkative that's okay too because you're just the chatty girl." One study in *Molecular Autism* concluded that "gender-based expectations" were one of two factors contributing to "the unequal gender ratio in ASD."

There is no doubt I got diagnosed at a younger age than the autistic women and girls I know. The very fact that clinicians were the first to point out to my mom that I needed more examination shows that even though the paradigm shifts about autism were fairly recent when I was a toddler, they were sufficient for the doctors to detect me. Perhaps this was because of my gender.

There are many possible reasons why autism is overlooked in girls, but one of the simplest is that practitioners are looking for the wrong symptoms. If girls do not conduct the same types of repetitive behaviors as boys or engage in them as frequently or if they do not focus on their interests as intensely as their male counterparts, diagnosticians might not recognize their behaviors as autistic. One 2014 study from the *Journal of the American Academy of Child and Adolescent Psychiatry* bore this out, finding that autistic females presented differently.

Girls may face more skepticism too. "It is also possible that high functioning females with ASD are differentially under-identified," the same study noted. Females in the survey were shown to have lower levels of "restricted interest," greater levels of social impairment, and "lower cognitive ability," along with greater social communication impairment.

To put it simply, the concept of being autistic is biased toward males. There is empirical evidence showing that autism is overlooked in girls, which leads to them getting diagnosed later in life. Women under eighteen in the Netherlands were diagnosed with Asperger's later than men under eighteen, according to a 2012 study published in the *Journal of Autism and Developmental Disorders* that surveyed 2,275 autistic people. For people over eighteen, the same was true when it came to autism diagnoses: women were diagnosed later. And autistic girls often get misdiagnosed. One 2010 study published in *Disability and Health Journal* surveyed 2,568 children born in 1994 who met the criteria for ASD around thirteen sites funded by the CDC. The study found that when an autistic girl presents as having an intellectual disability, "an [autism spectrum disorder] diagnosis may not be considered the primary problem and [intellectual disability] may be diagnosed, instead" and that girls, "especially those without a cognitive impairment," might be identified as having ASD later than boys. Put simply, if autistic girls with intellectual disabilities have the latter identified, their autism is more likely to be overlooked by clinicians. Either way, the diagnostic criteria is biased against them.

There is also evidence that many autistic women are better able to blend into their neurotypical surroundings, which allows them to go undetected. Men also "camouflage," but according to a 2016 study in *Autism,* women had higher camouflaging scores than men.

"When you go to get diagnosed, there are few specialists who have worked with girls, women, and LGBT folks," Yergeau said. "What clinicians see is the masking. They assume that you're nonautistic by virtue of the fact that you've developed a compensating or a passing behavior."

For many autistic women, "wearing a mask" is a coping mechanism, a way to hide their differences. However, it can ultimately have an adverse effect, as it prevents them from getting a diagnosis and proper accommodations along with it. Liane Holliday Willey encapsulated this phenomenon perfectly in the title of one of her books: *Pretending to Be Normal.* When I first spoke to her, in 2015, she talked about why it was easier for autistic women to mask compared to autistic men.

"We blend in easier because society makes excuses for our wanting to be alone or not wanting to go to a dance," she said. Autistic women and girls are often written off as simply shy or awkward.

Holliday Willey said she didn't exhibit any behavioral problems in school, and she was misdiagnosed with attention-deficit disorder and obsessive-compulsive disorder. "But behind the scenes, I was throwing rocks and breaking windows, and I was burning things, and I was cutting myself, and I was doing those hidden things that I think girls take out on themselves internally," she said. She wasn't diagnosed until many years later, after one of her daughters was.

This "taking things out internally" manifests in many ways. Ashia Ray, an autistic biracial mother who lives in Boston, went undiagnosed until she was thirty, when she was pregnant with her first son. She grew up in a predominantly white city, and behaviors that would typically be coded as autistic were dismissed as idiosyncrasies from her Chinese heritage or the product of being raised by a single mother.

Without a diagnosis, Ray said she was forced to say yes to everyone and act accommodating and compliant, much as neurotypical women are conditioned to do. Our society is fundamentally patriarchal and thus values women who are collegial and not too disruptive. Women who say no, who don't cater to others, or who "get in the way" of men are often labeled as frosty, schoolmarmish, or killjoys.

"It's interesting to see how, because I grew up undiagnosed, I was forced to kind of accommodate everyone else and change my personality to fit in," Ray told me. "Whereas you see a lot of white men who are diagnosed very young, and then they're given all sorts of excuses for bad behavior."

Ray said she got used to being uncomfortable, and because she didn't know she was autistic, she assumed everybody else was too. "When I realized I was autistic, I was like, 'Oh, you guys are not in constant pain? You guys don't say yes even though you want to say no?'" Ray said that her diagnosis enabled her to set boundaries, which autistic people are rarely taught about. "We are allowed to set boundaries and create some personal safety for ourselves," she explained. So many times, autistic people, particularly autistic women and nonbinary people, are victims not only of the skewed expectations of who can be autistic but of the standards that culture sets for their assigned gender.

Eryn Star also attempted to mask their autism, mostly as a means of survival.

"My mom pretty much viewed me being autistic as her fault," Star told me, explaining that their mother believed vaccines were the cause of their condition. As a result, Star became so adept at masking their autism that their mother believed Star had "recovered."

But the weight of presenting as neurotypical was heavy. "I think my brain was to a point where it thought the passing version of me was my actual self and it really wasn't," they told me. "I didn't really know what the autistic version of myself was."

Their attempts at masking, combined with the abuse by their teacher, was the final straw for Star, who developed generalized anxiety disorder. Then they saw a therapist, which finally turned things around for them.

"The therapist was like, 'You should embrace being autistic,' and I had never heard those words in a sentence because I was taught to hate it by my family," they said.

Autistic women whose neurotypes are ignored can face harsh consequences, just like other people who aren't recognized for who they truly are. One 2016 study published in the *Journal of Autism and Developmental Disorders* surveyed fourteen women who were diagnosed with autism later in life and found that almost all "experienced one or more mental health difficulty, with anxiety, depression and eating disorder being the most commonly reported." Many of the women were dismissed by professionals when they expressed that they thought they might be autistic.

And unfortunately, rates of sexual abuse are much higher among autistic women; the same 2016 study noted that more than half the autistic women surveyed experienced sexual abuse, most from within their own relationships. These women spoke about "feeling obliged" or "gradually being pestered" to have sex and thinking it was something required of them. Autistic women's uncertainty about social rules, such as not knowing they could say no if they wanted to refuse sex, increased their risk. This is not meant to victim-blame autistic women but rather to point out the need for awareness surrounding issues of sex and consent in the autism community.

Holliday Willey understands this confusion firsthand, noting that she complied with what men wanted when she was in high school. "Whatever they told me, I did. It could be from sexual [acts] to 'go write my paper for me.' [I thought that] this is what you do when you're dating," she said.

During our interview, Holliday Willey told me about multiple instances of intimate-partner violence that she experienced throughout her life. She explained the whirlwind of emotions that accompanied each incident, from embarrassment to shame to anger. Holliday Willey told me that she is only willing to discuss being sexually assaulted now because she has three daughters.

"And by God, I don't want it to happen to kids that come after me —any kind of kid, on the spectrum, not on the spectrum, trying to fit in, low self-esteem, whatever," she told me. The high rate of sexual assault on autistic women is the direct consequence of many of the misperceptions about autism—that it exists only in men, for instance, or that autistic people are asexual, thus rendering sex education irrelevant. On top of that, across the broader population, there is a general disregard for how sexual-assault survivors are treated. To move forward requires our society to reevaluate who we think can be autistic while letting go of our fears about teaching consent to neurodivergent people. When we are afraid of autistic people having a sex life, we make it easier for predators or abusive partners to hurt them, and we prevent them from learning that they have autonomy and the right to personal space. As mentioned in chapter 6, teaching autistic people about consent is vital to ensuring their safety.

Gender Identity

Autistic women and nonbinary people have sometimes struggled with how society tells them they're supposed to act. Some autistic women felt pressured to adopt traditional gender roles (and the burdens that come with them), such as wife, mother, and girlfriend, finding "this incompatible with how they wanted to live," according to the 2016 *Journal of Autism and Developmental Disorders* survey of fourteen autistic women.

"The schema of the autistic male is sort of quirky and awkward," said autistic woman Kris Harrison. But women aren't given the same liberty to be gauche. "If I miss a social cue, it's like I'm falling down on the job of being [an] emotional caregiver that [society] expects of most women."

Harrison is a statuesque blonde who wears flowing dresses accented by ornate jewelry that is perfect for stimming. If she comes off as a mix of eccentric professor and quick-witted mother, that's because she is. She's a professor in the University of Michigan's Department of Communication and Media, and two of her three children are also autistic.

Harrison said she has experienced confusion about her gender since she was a toddler. "A lot of the difficulty I had socially was over wanting to do and say things that I knew would be fine if a boy did and said them—because it was fine when my brothers did them—but if I did them it wasn't," she told me. For example, she recalled wanting to joke about a character in the Monty Python movie *Life of Brian,* "Biggus Dickus." "That realm of humor was off-limits," she said.

"I don't have gender dysphoria to the extent that I feel like I'd be more myself in a man's body," she told me. "I want to be able to talk like myself in this body and not have people go, 'How dare you!'"

Many autistic women have trouble bonding with female peer groups. In that same 2016 study, eight of the fourteen women surveyed discussed their difficulties trying to form friendships with neurotypical women. They said they "often found it hard to manage what they perceived were socially expected skills of a woman." This tracks for Harrison, who said she can be blunter in the company of men.

"A lot of autistic women will say that they're pretty comfortable around men," she told me. "I really kind of walk on eggshells in female collectives because you want to make sure your presence doesn't stress other people out."

Despite all the stereotypes of autism being a boys' club, I often say (only half jokingly) that I am usually in the minority, given how many women and LGBTQ people I encounter in the autism community. For instance, at Autspace, many attendees used "they/them" pronouns or were genderqueer, and two transgender young men gave an entire presentation dedicated to exploring how gender and autism intersect.

One of those presenters was Charlie Garcia-Spiegel, who spoke with me about the overlapping Venn diagram between the autistic and queer communities. According to Garcia-Spiegel, autistic people often don't pay attention to the same set of societal norms as everyone else, and with that freedom comes a vision. "We can see that a lot of the social rules around gender are"—he paused, trying to find a way to put his thoughts delicately—"bullshit, basically."

And research supports the idea that a large swath of genderqueer people are also autistic. In 2014, a survey in the *Archives of Sexual Behavior* showed that "gender variance was 7.59 times more common in participants with ASD than in a large non-referred comparison group." *Gender variance* is defined as "an umbrella term used to describe gender identity, expression, or behavior that falls outside of culturally defined norms associated with a specific gender," according to *Pediatric Annals.* Another article published in *LGBT Health* in 2019 found that children who were diagnosed as autistic were four times more likely to experience gender dysphoria.

"When we're forcibly distanced from social rules anyways, a lot of us kind of look at them and see, 'Oh, these social rules shouldn't really have an impact on how I carry myself through the world, and what my relationship to my body is,'" Garcia-Spiegel said. The large contingent of transgender autistic people is like the large amount of gay autistic people (to say nothing of autistic people who are queer and transgender): discovering one's gender identity can offer a road map to understanding one's autism. Learning that they are autistic can show people that they are not wrong for living outside prescribed social rules and norms, including ones for gender and sexuality. Once they accept that they are autistic, they realize that a lot of social norms are constrictive and should be questioned

And yet, being transgender or queer is not entirely liberating for autistic people who identify as such. Often, they encounter just as much ignorance as there is in the cisgender world.

This was true for Garcia-Spiegel. When he took a course on queerness around the world in his first year at Sarah Lawrence College, he struggled academically because he was not getting the proper accommodations. When he told his professor that, "He basically just implied that I wasn't qualified to be at college," Garcia-Spiegel said, even though he'd taken three Advanced Placement classes in high school and gotten the highest score possible in all of them. "It was just like this whole cycle of kind of not being taken seriously as a trans autistic person in a queer space because [I am] autistic."

Autistic transgender people face overwhelming discrimination and are often the subject of absurd myths. When J. K. Rowling, author of the Harry Potter books, began commenting extensively against transgender activism, she wrote that in the past decade, there had been a "4400% increase in girls being referred for transitioning treatment" and that "autistic girls are hugely overrepresented in their numbers," which is to say, transgender boys and men who were assigned female at birth. Rowling's comments essentially implied that transgender autistic people must have been deceived or given wrong information to want to transition. But Kris Guin, an autistic transgender man, wrote that Rowling's comments were a noxious mix of transphobia and ableism because they implied autistic people were incompetent and incapable of making their own decisions.

Some researchers have speculated—certainly less malevolently than Rowling—that just as increased levels of testosterone in the womb might lead to being autistic, it also might contribute to less common sexual identities and gender identities, as *Spectrum News* reported in 2020. But Jeroen Dewinter, a researcher at Tilburg University in the Netherlands, told the outlet that prenatal testosterone did not explain why autistic people assigned male at birth might present more feminine qualities.

Few people listen to autistic transgender people or ask them their reasons for transitioning. Their dual identities are not competing; they are complementary. The acceptance of each affords transgender autistic people new freedoms they otherwise would not have. A lot of the bias against this population is also rooted in the idea that autistic people cannot understand what is in their own best interests. This pernicious ableism compounded with transphobia implies that autistic people cannot understand their own gender identity.

Still, autistic people know what they want and need. They are the ones who know best about their identities and how to ensure that their bodies line up with what is in their minds. The only thing they need from other people is affirmation and support.

"A Girl Like Me"

There is still plenty of work to do to close the gender gap in the autistic community, though in recent years, significant strides have been made increasing the visibility of autistic women and girls. The DC-based Autistic Women and Nonbinary Network services autistic people across the gender spectrum and it has released books like *What Every Autistic Girl Wishes Her Parents Knew.* Furthermore, television portrayals of autism are now broadening to include autistic girls as well as boys.

Australian stand-up comedian Hannah Gadsby discussed being autistic in her second major stand-up special *Douglas* and said learning she was autistic allowed her to "be kinder to myself." Kayla Cromer, who is autistic herself, played an autistic character named Matilda in the show *Everything's Gonna Be Okay.* Cromer came out as autistic during a press event for the show in 2019. She admitted that for years she didn't think she was funny. But her acting coach said she had quirks that resembled Sheldon Cooper, a character on the television show *The Big Bang Theory* who is often speculated to be autistic. (Though the show's creators object to the label, I've always seen the character as a reductive portrayal that is the autistic equivalent of blackface minstrelsy. Sheldon perpetuates the stereotype of a geek savant whose eccentric behaviors and rudeness toward other people—particularly women—are either a source of comedy or excused by his genius.)

Fortunately, representation of all types of autistic people is improving because we're now recognized as gatekeepers for telling our own stories. When Pixar Animation Studios released the short film *Loop* in 2020, they were careful to include autistic voices when developing the neurodivergent protagonist. The movie takes place in a summer camp where Renee, a nonspeaking autistic girl of color, and Marcus, a neurotypical African American boy, must navigate a lake in a canoe. Erica Milsom, the film's director, told me that in the initial storyboards, Renee was white. But when she consulted with the Autistic Self Advocacy Network, they suggested Renee be a person of color to bring awareness to the fact that autism occurs across all races.

The decision to make Renee be both autistic and a girl was incidental, Milsom added. All she knew was that she wanted to have a nonspeaking character represented in the movie because she knew similar people in her life.

"I was like, 'Okay, well, I need to have a specific reason why this person doesn't talk,'" Milsom said. This led to her asking friends who have children who communicate in different ways. "As we went through and talked together, people started talking about autism and all the layers of sensory experience that differ along with communication and language," Milsom said. "I just didn't know that stuff, and I'm like, 'This is amazing. This is such an interesting, deep identity and point of view.'"

To me, the effect of autistic people's input was clear in *Loop*. Throughout the first part of the film, Renee plays the ringtone on her phone to self-soothe. Any autistic person who likes hearing the same sound, whether it's the same song played on a loop or watching the same movie over and over, could absolutely relate. Showing a non-verbal autistic person using a ringtone to make herself feel better is significant, since too often, the things autistic people use to cope with the world are pathologized. It's especially important to use a female character of color to show this, as it broadens the scope of who people think can be autistic.

This is a big shift from the film *Rain Man*, whose consultants, Bernard Rimland and Darold Treffert, insisted that the movie had to end with Dustin Hoffman's Raymond Babbitt character returning to an institution. Other programs with autistic characters, like ABC's *The Good Doctor,* have hired consultants, but it's unclear if they are autistic or not. The popular children's TV show *Sesame Street* worked with both Autism Speaks and ASAN to create Julia, its first-ever autistic character, although ASAN eventually parted ways with the show because its members felt the marketing around Julia further stigmatized autism and autistic people. Milsom's consultation with an autistic-led organization for *Loop* is truly progressive, as it allowed autistic people to have the strong creative input to make sure Renee accurately portrayed autism.

I also noticed the influence of autistic voices in *Loop* when the movie showed what happens when autistic people experience a meltdown. Milsom said she spent two and a half months researching this scene by watching videos of meltdowns on YouTube.

"I feel like if you're writing, you need to know the interior life of the character," she said, and as a result, she wanted to know not only what a meltdown felt like, but also what preceded it and what was needed in the moment. Meltdowns are, of course, difficult but also a normal part of the autistic experience.

Milsom is grateful for autistic people who tell their stories. She consulted with autistic friends and organizations while creating the film, and she even cast Madison Bandy, a nonspeaking autistic girl, to play Renee. To make her as comfortable as possible, the audition was held at Bandy's art center and her vocal parts were recorded at her house.

The creation of fictional autistic female characters that are played by autistic people coincides with the fact that autistic women and girls are becoming public figures in areas only tangentially related to autism. Greta Thunberg, the teenage Swedish climate activist, is perhaps the most famous autistic person in the world. Her supporters have suggested she deserves the Nobel Prize. Just like Eryn Star, Thunberg was born a generation after me, which means she had the good fortune of growing up in a society that at least had a working knowledge of autism. This might be why Thunberg sees being on the spectrum as a reason she can be so strident in her advocacy on climate change. But Thunberg received a diagnosis only because learning about climate change in primary school sent her into a deep depression. She ate little, lost a tremendous amount of weight, and stopped speaking.

Thunberg's story—both her triumphs and the fact that it wasn't until she was eleven that she was diagnosed—show the progress that's been made but also highlight how painstakingly slow the paradigm shifts for autistic women and girls are. The same can be said for all femme-presenting people and queer people. And it's even tougher for people who, unlike Thunberg, don't have supportive parents who are willing to help and accept them. Our misunderstandings about autism are rooted in the idea it affects only boys, and that misperception intersects with the expectations our society has about women. Before we can address the real gap in diagnosis and services for autistic women and LGBTQ people, there first needs to be an acknowledgment that they are a vital part of this community.

8

"Say It Loud"

RACE

No news incident shook me to my core more than the 2016 shooting of Charles Kinsey in north Miami. Kinsey, who is Black, was the behavioral aide of Arnaldo Rios-Soto, a Latino autistic man who had wandered away from the home where Kinsey was caring for him.

When police found Rios-Soto sitting in the middle of a road, they thought he was armed, although he only had a toy truck. In video that captured the shooting, you can see Kinsey try to de-escalate the encounter while also trying to assist Rios-Soto. When the police shot Kinsey, he was lying on the ground with both hands in the air. Despite being left on the ground bleeding for twenty minutes without medical attention, Kinsey survived.

Still, Kinsey's shooting devastated me. Had he not been taking care of Rios-Soto, he likely would not have been shot. At the same time, as a Black man, he likely knew his body was always vulnerable in every police interaction, but he still attempted to save his charge. But in the end, he wound up another Black victim of police violence, though he thankfully lived. The fact that he was shot while police were responding to a Latino man was terrifying because it meant that those who were even in proximity to autistic people were also not safe. I knew that Black and brown people were more vulnerable to police violence. I had graduated from college a few months before police officer Darren Wilson shot Michael Brown in Ferguson, Missouri, which led to widespread protests and cemented the chant of "Black lives matter" in the public consciousness. Only a few days before the Kinsey shooting in north Miami, Philando Castile, a Black man in Minnesota, was shot and killed by a Hispanic police officer. But until Kinsey's shooting, I hadn't considered how autism factored into the epidemic of police brutality. I didn't think that autism, which sometimes makes it difficult for me to make eye contact or requires stimming to calm down, could be seen as a threat. My fear increased when I read that the local policemen's union justified the shooting by saying Rios-Soto, not Kinsey, was the target. There is this perception that autistic people are permanently children and angelic, but that is largely because the common archetype of an autistic person is a white person.

This "perpetual child" image is the result of the fact that most autistic people featured in the media are white. America largely assumes innocence (and excuses fault) for many white people, and often in the case of autistic people, society chalks up their mistakes to their disability. Autistic people of color—be it Black or brown—aren't given that luxury. We aren't given the benefit of the doubt, and our odd behaviors—the way we rock, the way we avoid eye contact, the way we stim to calm ourselves down when around police—become cause for suspicion. All of this stems from the perception that autism is a white condition; a racialized autism means that Black and brown people on the spectrum are overlooked by clinicians while their behavior is perceived as dangerous by the police and the broader public. A lack of focus on autistic people of color is the difference between innocence or guilt in the eyes of the law and the world.

Reading how the police saw Rios-Soto as a threat showed me the exact difficulty that comes with navigating being autistic in a nonwhite body—your neurotype is not something that can be known, so the wrong call or movements can be deadly.

"American Skin"

A. J. Link is an affable and inviting person. When I showed up at his doorstep a few days after Christmas in 2019, he welcomed me into the immaculate apartment he shares with his roommates and his dogs.

Link is roughly my height (I'm around five eight) and was wearing a sweater and sweatpants. He sports a full beard that I'm envious of, given my inability to grow a proper one. Link was finishing his third year of law school at George Washington University (something that you would never have guessed from his laid-back attitude) and I was here to interview him because he founded the school's Atypical Student Society to assist neurodivergent students.

Link told me that the George Washington University law school prided itself on being diverse and inclusive, "but something that gets lost in diversity is neurodiversity." While almost 10 percent of the student body at the law school is registered with its disability support services, the real number of students in need of support is probably double that because of the stigma attached to disability.

"No one wants to be looked at as 'other' or 'special' or 'handicap' or whatever," he said. "It's tough and frustrating, but I hope that we can teach people that neurodivergent individuals are just different, not less."

Sitting in Link's living room as Liverpool Football club plays in the background, I was suddenly reassured by this scene: me, a Latino journalist, interviewing Link, an African American autistic advocate at a top-tier law school. Neither of us fit the stereotype of what an autistic person looked like, yet here we were.

The outdated idea that autism affects only white people goes back to the earliest studies about the condition. Again, we must return to Leo Kanner's initial 1943 study. In "Autistic Disturbances of Affective Contact," of the eleven children in his survey, none were people of color. Nine of them were from Anglo-Saxon families and the other two were Jewish. He later said "all but three of the families" of most of the children studied were represented "either in *Who's Who in America* or *American Men of Science,* or in both," implying that autism affected only the highest socioeconomic classes. In a cruel irony shown in a 2002 documentary entitled *Refrigerator Mothers,* one Black mother named Dorothy Groomer was told by clinicians that her son Stephen couldn't be autistic and she couldn't even be considered one of those cold and unloving parents as described by Bruno Bettelheim.

Kanner's idea that autism affects mostly affluent, upper-middle-class families persists. This comes at the expense of autistic people of color, women, and gender nonbinary people who are often overlooked, misdiagnosed, and fail to get the services that they would otherwise receive. A 2007 study from the *Journal of Autism and Developmental Disorders* found that African American children "had 2.6 times the odds of receiving some other diagnosis" compared to white children. Over time, there has been some effort to reduce the diagnosis gap. In 2020, the Centers for Disease Control and Prevention released a report about the state of autism in 2016 that said that the gap between white children diagnosed with autism compared with Black and Hispanic children was narrowing and that for the first time, the Autism and Developmental Disabilities Monitoring Network, which tracks the number and traits of autistic people, found no difference between the number of Black and white children identified with ASD by age eight. But still, white and Black children are 1.2 times more likely to be diagnosed than Hispanic children. That the gap has narrowed is welcome news indeed, but the fact that it took until 2016 to close the gap between Black and white autistic kids demonstrates that the clinicians who developed the diagnostic criteria for autism did not have the former in mind.

Even when accounting for all socioeconomic factors that might mitigate the chances of Black children receiving the same diagnosis as white children, Black children still face major hurdles to receiving the diagnosis. A 2002 study in the *Journal of the American Academy of Child and Adolescent Psychiatry* surveyed 406 children who were eligible for Medicaid in Philadelphia and received services for autism between 1993 and 1999. It found that white children were diagnosed at 6.3 years of age while Black children were diagnosed at 7.9 years of age, and they required more time in treatment before they were diagnosed. Children who receive Medicaid are usually from poor or low-income families or are disabled, so this was as close to a symmetric comparison of white and Black children as possible. Years later, the CDC confirmed that Black and Hispanic children get diagnosed later than white children.

Link, for example, wasn't diagnosed until his early twenties, despite the fact that his parents had him tested in childhood. He grew up in Polk County, Florida, where he was one of the few Black kids at a predominantly white school. "I think that stopped people from going deeper into why I was weird and different," he said.

Link said he was finally diagnosed in college because his therapist suggested it. He was struggling at the University of South Florida. "I would often skip exams or not turn in papers, or turn papers in late, which I still do, but at least I know it's because I have executive-functioning issues and that's just the way it is."

He spoke with his therapist about why he felt different from his peers, which led to his diagnosis of what was then labeled Asperger's syndrome. After a month of testing, it was confirmed he was on the spectrum. "It made sense. Being told that I was autistic made my life easier in a way. I didn't have to force myself to mask, I didn't have to pretend like I was enjoying what everyone else was enjoying," he said.

Link said he doesn't know for sure how his life would be different if he were white, but he thinks that he would have received his diagnosis much sooner rather than being labeled as a problem child with behavioral issues.

Unfortunately, this is a common problem for Black autistic children. According to the 2006 *Journal of Autism and Developmental Disorders* study, Black children were 5.1 times more likely to be diagnosed with adjustment disorder—which is characterized by stress, sadness, or hopelessness, or physical symptoms from a stressful experience—than with ADHD. Black children are also 2.4 times more likely to receive a diagnosis of conduct disorder—wherein children have persistent behavioral and emotional problems characterized by "difficulty following rules, respecting the rights of others, showing empathy, and behaving in a socially acceptable way"—than a diagnosis of ADHD. The same study also noted that clinicians might also diagnose Black children with oppositional defiant disorder.

Finn Gardiner, who is Black and autistic, was diagnosed when he was around two or three, but he still faced extraordinary disciplinary actions in elementary school when he had meltdowns because of stress or bullying.

"I feel that I experienced a lot of disproportionate disciplines, and I feel if I were white and autistic, I don't think I would have been excluded from class as much," he said. At the same time, Gardiner was in the gifted and talented program throughout much of elementary school and middle school. "But even with that label I was treated as a troublemaker, you know, because of my race and disability," Gardiner said. "I'm pretty sure if I were white and I had a gifted and talented label and I was complaining [about] the lack of challenge and being bullied, they would have reacted very differently; they would have been sympathetic."

Link's and Gardiner's stories diverge in many ways; Link wasn't diagnosed until college and Gardiner was diagnosed as a toddler. One is a lawyer and the other is a researcher. But the fact there was an immediate impulse to label them both as problem children rather than accommodate them shows how racist ideas about Black people permeate how autistic Black people are treated. It shows there is an impulse to punish them for certain behaviors rather than seek to understand the cause of their actions. Autistic children are given little empathy to start, but Blackness reduces any small ounce of sympathy they otherwise might receive.

Sara Luterman, an autistic writer and journalist based in Takoma, Maryland, said that the fact that she is white prevented her from severe punishment in school, noting that she was suspended rather than expelled when she threw a chair at her teacher.

"I got in fights with other kids a lot. But I was also very small, and very white, and very female," she said. "It's like a teeny-tiny chair, and I can't throw it that hard, so nobody actually got hurt. And I think that made a difference. And also, just there's this perception that I'm not threatening. I'm just easily upset."

"I Contain Multitudes"

The fact that autism is seen as a "white" condition affects who white people perceive as autistic and therefore worthy of accommodation. But there is evidence that Black parents perceive symptoms of autism differently from white parents. A 2017 study in *Autism* found that "Black parents reported fewer concerns about their child's ASD-specific behaviors than White parents." But Black parents' oversight wasn't because they didn't care about their autistic children. The same study said that "cultural norms surrounding discussion of ASD within Black communities may also impact parents' knowledge about the disorder" and that "parents of different cultural backgrounds may also have different expectations or interpretations of a child's behavior, which may affect their report of concerns." This is to say that it is possible that different cultures may have different values, which leads to parents emphasizing different aspects of their children's behavior and development.

The discussions around police violence in the United States helped the term *intersectionality* enter the public consciousness. While the term is frequently thrown around in political discourse and social media, it was defined by Black feminist legal scholar Kimberlé Crenshaw in 1989 in a legal paper. Essentially, *intersectionality* means that people's multiple identities often create different experiences of discrimination and privilege than those who might occupy one marginalized identity. In Crenshaw's text, she said that Black women experienced discrimination that was similar to both Black men and white women but also unique because of how these identities intersected.

"Often [Black women] experience double-discrimination—the combined effects of practices which discriminate on the basis of race, and on the basis of sex," Crenshaw wrote in 1989. "And sometimes, they experience discrimination as Black women—not the sum of race and sex discrimination, but as Black women."

Since then, Crenshaw's ideas about intersectionality have extended to LGBTQ people, mixed-race people, nonbinary people, and, of course, disabled and autistic people. Just as Black women experience double discrimination of racism and sexism, autistic people of color encounter bias that is unique to their combination of identities. This is not to create a contest of who is more oppressed but rather to understand how certain people's lived experiences are unique so that the discrimination they face can be addressed.

Morénike Giwa Onaiwu, the chairwoman of the Autistic Women and Nonbinary Network's Committee on Autism and Ethnicity, spoke to me about her own intersectional identity. She is the child of West African immigrants and talked about how Black Americans and people of color often worry that "my own people don't understand me." Autistic people of color find themselves trying to find ways to participate in their own culture while also feeling completely out of place. While this is surely something all autistic people struggle with because of the cultural expectation that one should "fit in" with one's own ethnicity, it can feel especially overwhelming for autistic people of color.

For example, Giwa Onaiwu told me that some of her children are also autistic, including her son, who she said loves to jump around but has little hand-eye coordination, which prevents him from being able to play football. "I live in Texas. And to Texans, football is life, you know what I mean?" she told me. "My child has never played football, and that's weird to people. That's odd. That's what you're supposed to do, what you're supposed to be able to do." Giwa Onaiwu said that growing up, she felt that when she moved around, she had to struggle to prove that she was sufficiently Black and was accused of "talking white."

"It was frustrating, 'cause I'm like, 'I'm just as Black as you. In fact, I'm Blacker. I'm African!'" she explained.

Timotheus Gordon (the graduate student whom I interviewed about accommodations in chapter 2) said he often felt a need to prove he was sufficiently Black and masculine. While he loved Japanese anime and *My Little Pony*, he also loved sports, which he used to connect with other men.

"The first way I got into the in-crowd with [neurotypical] peers is showing people that I can play sports," he said. "I never was gonna be the next Lebron James or Marshawn Lynch, but I showed people that I know the rules, I know how to follow the rules."

Growing up as a Black man in Chicago, Gordon said he felt like he had to show he was sufficiently into the hip-hop and rap music that was popular at the time. "I had to dress a certain way, behave a particular way," he said, which was difficult for him at first, but he learned to "fit into that culture" when he was in high school.

Gordon said when he first met other autistic people, he found it difficult because he had spent so long trying to blend in with his peers. This experience is typical for autistic people of color who wind up having to navigate multiple spheres to prove they belong. They have to prove to educators and physicians that their symptoms are sufficient to warrant an autism diagnosis and receive the necessary services, but at the same time, they also have to confront the fact that they're not represented in the autistic community and so they struggle to fit into spaces that were largely built by white people without them in mind.

What Do I Do with This Brown Skin and Autistic Brain?

But it is not just African Americans who are overlooked when it comes to autism diagnoses. One of the biggest barriers to diagnosis is the existence of a language gap, so it can be especially challenging for immigrants to get the support and services they need.

As a third-generation American, I often worry that I am too far removed from my family's roots in Mexico, but the cruel irony is that I likely received my diagnosis *because* my family speaks English as our primary language and we are largely assimilated. This does not make us better than those who have not assimilated as easily; assimilation is often forced on immigrants, the children of immigrants, and their grandchildren. It's the only way to have the opportunity to achieve a modicum of success in America, where white English-speaking people write most of the rules for the economy, politics, and culture. Sometimes, the capacity to assimilate means the ability to access the tools to learn English and gain exposure to what is acceptable. Assimilation is capital, and those without the ability to assimilate often can't give their autistic children the chance to be accepted as autistic and therefore have their humanity accepted.

A study published in 2013 in *Pediatrics* found that in a sample of 267 primary care pediatricians in California, 30.4 percent offered general developmental screenings and 42.9 percent offered screenings for autism spectrum disorders, but only 17.7 percent offered general developmental screenings in Spanish and only 28.7 percent offered screenings for autism spectrum disorders in Spanish.

Katharine Zuckerman, an associate professor at Oregon Health and Science University who coauthored the study, said the irony is that people are often assessed for autism because they have trouble communicating with others. "So, if you go in to have your communication assessed by someone who doesn't even speak the language that your family speaks, [and] they use an interpreter, it's such a confusing situation for kids," she said, adding that on top of that, many assessment instruments aren't even validated in Spanish.

"We're using tools that we don't know how good they are, usually used by people who aren't doing it in the appropriate language," she told me. "And that's even if you get to having a diagnosis, which a lot of families don't get to. It's so messed up in a lot of ways."

Zuckerman told me that there had yet to be a study comparing Latino diagnosis rates to white diagnosis rates. She said that, anecdotally, it is understood that there is a higher threshold for autistic Latinos to get diagnosed compared to white kids who fall on roughly the same spot of the autism spectrum.

"Those kids who aren't being diagnosed with autism are being diagnosed with nothing, or they're being diagnosed with something else," she told me. "These kids are almost functioning okay, and the downside of that is those are the kids that actually benefit the most from therapy."

Arianne Garcia (no relation to the author) is based in my dad's hometown of San Antonio, Texas. She told me she did not get diagnosed until she was twenty-five.

"There's a reason I didn't get that diagnosis when I was younger," she said. "It's because I don't come off that way. I'm different, and my experiences as a minority have shaped my social responses. If somebody is telling me something in a rude tone, I automatically think, Is it because I'm brown? Is it because I'm a woman? Really, it could just be that I'm awkward and I don't realize it, or they don't understand me, or I'm throwing them off guard."

Listening to Garcia's experience, I couldn't help but be reminded of my own path to getting diagnosed. My mom noticed that I wasn't developing in a normal manner; I had trouble nursing, didn't walk until I was eighteen months old, and hated when the phone rang or when announcements were made in the grocery stores. When my family lived in Wisconsin, my mom read a newspaper article about autism screenings for preschool and kindergartners. She took me and my sister in, but initially, evaluators wanted to send me on my way. It was only after taking a second look at the results that they asked my mom to bring me back for more tests. It would be easy for me to say that race played a role, but more likely, the criteria were still shifting (this was around the time that Asperger's syndrome was added to the *DSM-IV*), and the evaluators didn't initially catch something they should have.

But midway through my evaluations, my dad got a job in Folsom, California, and my family moved halfway across the country. While we were there, my mom said I did not exhibit the stereotypical behavioral problems that cause so many parents to somewhat callously describe their children as menaces. My mom said I didn't exhibit signs of aggression, but I did like to have pressure applied to me, so I would lie on the ground underneath a beanbag chair to feel that weight and pressure.

My mom said that the school system kept saying I was "fine" because administrators did not want to spend the money on services. If you ask my mom, she also thinks that I was overlooked was because we were Mexican American. The only reason I finally wound up with my diagnosis was due to a coincidence: My dad's new boss was married to the director of the special education programs for Sacramento County. Once my mom talked to her, I had no problem getting accommodations in school and was shortly thereafter diagnosed with Asperger's syndrome.

The fact that I got a diagnosis thanks to my mom's relentless advocacy makes me feel grateful, but it also makes me shudder because I know that far too many autistic people don't have the same chance I did. As my mom said, "Unbeknownst to us, we were pulled into a current that would prove to be to your great advantage." Everything had to go right, from my mom reading about the screenings in the newspaper to the evaluators being willing to take a second look to my mom getting a director of the system on our side. Furthermore, my mother was able to be a stay-at-home mom for the first twelve years of my life before my parents divorced. That enabled her to obtain as much information about my condition as possible. Both of my parents spoke English as their first language, which allowed them to access information. As another study Zuckerman coauthored found, the biggest barrier to Latinos with limited English proficiency was "parent knowledge about ASD" and "trust in providers." I didn't have to worry about either.

The biggest barrier that non-Latino white families faced was stress during the diagnostic process. Latinos with English proficiency, like mine, were similar to non-Latino whites overall, but the stress of the process was less common. Zuckerman told me that the findings don't mean that Latino families didn't experience stress during the diagnostic process; they did, as did all families in the survey. But if parents don't understand what autism is, how to get help for their child, or what comes next, the process is that much more difficult.

"I had a parent once tell me, 'When the doctor told me my child had autism, I thought it was cancer and that he was going to die,'" Zuckerman said. "I'm not saying most Latino parents think that, but Latino parents don't know as much about autism and so that in itself is a huge marker of stress."

Staying Alive Becomes an Act of Courage

Autistic people of color recognize that as we try to navigate our identities, others simply identify us by our skin color, which can lead to dangerous or deadly consequences. In 2020, America—and the world—was shocked and horrified when a Minneapolis police officer killed George Floyd, a Black man, by placing a knee directly on his neck. In March of that year, Breonna Taylor, an emergency medical technician, was shot and killed by police officers who had barged into her apartment in Louisville, Kentucky. There were widespread calls for the arrests of the officers who killed her, and that became a rallying cry for many and on social media. The fact that video was captured of George Floyd's murder and, later, Jacob Blake's shooting in Kenosha, Wisconsin, seemed to finally make white Americans see that their Black countrymen and women were not being hyperbolic when they said they feared police. The protests, demonstrations, and occasional damaged property triggered President Trump to tear-gas demonstrators in front of the White House but also seemed to shift national opinion in favor of Black Lives Matter, which was not the case when the phrase entered the public lexicon in 2014 (though support among white Americans began to decline after a summer of protests).

But in this shift in public opinion, it is essential for the public to remember how vulnerable Black, Latino, and otherwise marginalized autistic people are to police violence. A 2017 study that appeared in the *Journal of Autism and Developmental Disorders* found that 20 percent of autistic people had been stopped and questioned by the police and 5 percent of them had been arrested by the age of twenty-one. And while there is little data on how race intersects with autism, there have been numerous troubling instances of autistic people of color being victimized by police.

In 2012, a Black autistic boy named Stephon Watts was killed by police in the Chicago suburb of Calumet City. He was only fifteen years old. The police officer said that Watts had lunged at him with a steak knife; Watts's parents say it was a butter knife. Both the Circuit Court for Cook County and the Illinois Court of Appeals ruled in favor of the police.

In Chicago in 2017, dashcam footage showed an off-duty police officer shooting eighteen-year-old Ricardo Hayes, an unarmed Black autistic man, who fortunately survived the encounter. The sad irony in this case is that Hayes's family had initially called the police for help after Hayes had gone missing. Unlike in Watts's case, Hayes was unarmed, and the City of Chicago agreed to pay $2.25 million to his family. However, the officer who shot him, Sergeant Khalil Muhammad, received only a six-month suspension.

Watts's and Hayes's stories differ in many ways. Police used the knife Watts was holding to justify their use of force, while Hayes was unarmed. The police officer who shot Hayes was reprimanded and there was a financial settlement, whereas the Watts family had no such closure. Watts died and Hayes survived. What unites them is that they were both young autistic Black men who were shot by police because they lived in a world where racialized autism was considered a threat.

There are many things that could make autistic people seem threatening to police. The fact that we stim and fidget could be mistaken for a sudden movement—like reaching for a weapon—that police view as a threat. The fact that many autistic people don't make eye contact could be seen as not being respectful to police officers. If police don't know that autistic people speak in echoing, they might interpret a person as indirect or evasive. All of these risk factors are combined with the fact that young Black boys are often perceived to be older than they are and are viewed as less innocent than their white counterparts, according to a 2014 study from the American Psychological Association. When I interviewed Ron Hampton, a Black former police officer in Washington, DC, who has an autistic son, for a piece after Charles Kinsey's shooting, he told me that "when our children have episodes, we call each other," referring to other people in the community rather than the police.

The increased attention to police violence against Black lives in 2020 also returned a focus on the August 2019 killing of Elijah McClain by police in Aurora, Colorado, for which the three white officers were initially cleared. McClain's family said he was stopped on his way home from getting iced tea for his brother. The Aurora Police Department said it had received a call about a suspicious person in a ski mask "acting weird" and "waving his arms around." Initially, McClain reportedly told officers "I'm an introvert" and asked them to respect his space. Police say McClain resisted arrest and one officer warns in in the video that McClain was going for one of their guns. Eventually, police put him in a carotid hold, which is meant to restrict blood to the brain. When first responders arrived, they injected him with the sedative ketamine. McClain went into cardiac arrest twice en route to the hospital and died on August 30.

It has not been disclosed whether McClain was autistic. But many autistic people and their loved ones immediately saw similarities between McClain and themselves, particularly because McClain's last words to police were "I'm just different." Autistic people and their families fear that their being "different" makes them a target, that it makes them seem suspicious or stand out. All autistic people, but particularly autistic people of color, know that we could be Elijah McClain and that telling police officers we are "different" will not save us.

This does not mean that police are inherently bad or evil; rather, police are often ill-equipped to handle situations with autistic people. In recent years, there has been a trend to train police to better interact with people on the spectrum, and there have been efforts to use virtual reality to train police. In 2017, a year after Kinsey's shooting, Florida passed legislation requiring autism training for police. The training, as laid out in the legislation, included guidance on how to recognize symptoms of autism and examples of appropriate responses when interacting with autistic people. Pennsylvania and New Jersey have passed similar bills.

Still, there won't be an easy fix to this gap in understanding. A 2016 study published in *Police Practice and Research* found that 23 percent of respondents to a survey said their agencies had not mandated autism training. Furthermore, though Illinois began to mandate training for new police officers in 2008, the Stephon Watts shooting happened four years after the law had been passed. The public deaths of Black people that disturbed the public consciousness led to a call for defunding police departments. Supporters of that movement make the case that psychiatric services or other types of intervention might be more effective at helping people with mental illness than police intervention. Whether one supports or opposes defunding the police, it is likely that finding alternatives to police could benefit mentally ill, autistic, or otherwise disabled people.

The fates of Watts, Rios-Soto, and Hayes reminded me of my own interactions with police. As a Boy Scout, I was taught to see the police as respected civil servants meant to protect the community. Some of my scoutmasters worked in sheriff's offices and some of my fellow Boy Scouts became police officers themselves. Meeting and personally knowing many police officers is precisely why I do not believe all police officers are bad, inherently racist, or have malicious intent. Most of the time, they are trying to do their jobs and go home, just like I am.

Growing up, I only interacted with the police when they busted my garage band after boring neighbors called them. Even then, I didn't see police officers as a threat. But my mom, who grew up in East Los Angeles, had a very different relationship with police. Like most parents of Black and Latino teenagers, she told me to be respectful of police if I was ever stopped. If I saw a police incident, she taught me not to wander over to see what was happening lest I wind up being apprehended myself. For a long time, I thought she was being a bit too hysterical; we were far removed from Los Angeles and lived in the suburbs, where we knew the police officers.

That would change in my junior year of college when one of my classes involved traveling to Raleigh to cover the state legislature. It was my first real taste of legislative reporting and it quickly became a favorite beat of mine. I would hitch a ride with my friend Kathryn to Raleigh and then cruise the legislative office buildings as well as the main building on Jones Street. It was exhilarating and would serve me well when I later covered Congress.

But one week, Kathryn could not take me to Raleigh, so I took a bus to Jones Street. In addition to writing a story for class, I often reported stories for the campus paper, the *Daily Tar Heel,* from the legislature. When I got there, I parked myself in the gallery and pulled up my laptop to report on the petty accusations of bad faith between the all-powerful Republican majority and the hapless Democratic minority. Soon, one of the attendants approached me and said he would need to see inside my backpack if I stayed inside the gallery. I decided to duck out and wait near the large doors on the bottom floor, where legislators would exit once the session was adjourned.

Shortly thereafter, two uniformed police officers approached me and sternly asked what I was doing. I explained I was a student at UNC, but I didn't have a press pass. I pulled out my student ID card and then my proper ID card, but since it was issued in California, it only raised their suspicions. As I began to panic, the things that went through my mind were that I was a kid without enough identification, I was at a government building in North Carolina, and I was Latino. Those statues that lionized the Confederacy that I usually ignored were now a startling reminder that I was a guest there.

The whole thing lasted five minutes tops and nobody else seemed to notice; all of the legislators exited the doors and went on about their days. But I was absolutely terrified that somehow, I would wind up going to jail. By that time, I had remembered that I still had my Arabic textbooks in my backpack (I was taking Arabic classes to fulfill my foreign-language requirement and hoped to one day become a foreign correspondent). I worried that those books, along with my California ID, dark skin, and lack of connections to the state, would make them think I was plotting a terrorist attack.

The officers asked if anyone could vouch for me. I called Daniel Wiser, my editor at the *DTH,* and handed the phone over to the officers. Eventually, they left me alone after explaining I was considered suspicious for not wanting to reveal the contents of my bag.

On the surface, no harm was done; I did my interviews, took the bus back to Chapel Hill, and finished my work for the day. It even became a joke in the *DTH* office that I was willing to go to jail to get a story. But for those brief moments, I was aware of what it means to be a person of color in America, particularly in the American South. I was frustrated, angry, and scared.

To this day I wonder what would have happened if I had started to exhibit autistic behaviors. What if I had started stimming to calm myself down, as I occasionally do when I am scared or frustrated? I probably would have become too upset to let the police know I was autistic, and if I had made one false move, they could have put cuffs on me or worse.

Giwa Onaiwu told me about an experience she had in 2018 when she was stopped by a police officer while driving when she was using a toy for stimming. "I realize [the police officer] thinks I have a weapon," she said. "I'm like, 'No, this is a stimming toy.' He was like, 'What do you need it for? What are you doing with it?'" She said the officer initially reached for his gun but luckily didn't pull it on her. "I'm just grateful that he didn't just pull his gun and shoot me because he saw something fast moving out of a Black woman's hand," she said. Still, the incident was unnerving.

These days, A. J. Link does not drive much since he lives in DC, but he said he has been pulled over by the police before. Just in case, he has a signifier on his license showing he is autistic. Link knows he is fortunate because he is a law student and has knowledge of his rights that other autistic people and people of color might not have.

Link and I had our talk right before his last semester of law school. By the time this book is published, he will have graduated and passed the bar so he can practice law. He has said he hopes to get involved in space policy, which is another similarity between us (when I was younger, I wanted to be an astronaut). To this day, when Link is having what he calls fits, he sometimes asks his friends to play Carl Sagan's famous *Pale Blue Dot* excerpt, in which the late astrophysicist and popularizer of science describes an image of Earth that shows it to be a small speck in a vast pitch-black void.

"If it's a really high-end incident, I'll put that on YouTube right quick, or I ask someone to put that on YouTube, listen to three or four minutes, and it kind of just puts me at ease," he told me.

In Sagan's famous work, he deduces from looking at this little fleck of blue on which all of humanity has ever lived that it is "our responsibility to deal more kindly with one another, and to preserve and cherish the pale blue dot, the only home we've ever known."

But autistic people often feel like we may be on this planet, but we are not of it. Temple Grandin, the famous animal scientist, once told Oliver Sacks in a *New Yorker* interview that that she struggled with complex social interactions and that many times, "I feel like an anthropologist on Mars."

Though Grandin has faced scrutiny by other autistic people for her sometimes outdated or myopic ideas about autism or autistic people, that feeling that we are on another planet is familiar. If we do not belong to this neurotypical planet, how do we compel others to be kind to us? We may come to understand the planet's culture and learn its language but speak it with an autistic accent. Furthermore, autistic people of color not only have to navigate this foreign planet but are thrust into this country's most original and fraught struggle. Some of our nation's greatest sins are its racist legacies. While some autistic white people can simply ignore this reality, autistic people of color cannot. Unlike actual Martians, this is our home planet; this skin that is our space suit can be a target for pain. So we wander through this space, hoping to find balance but knowing we will be pulled by these dual gravities of our neurotype and skin color. Autistic people like Link, Giwa Onaiwu, Gardner, Gordon, Garcia, and Garcia-Spiegel, and me will always live with that.

"Living in America"

As a third-generation Mexican American, I grew up in largely suburban settings. Even when my family moved to the more heterogeneous Southern California, it was to Chino Hills, an idyllic town that could easily be in the Midwest (a far cry from East Los Angeles, where my mom was raised). When I am around other Latinos, I am often ashamed that I don't speak fluent Spanish, even though a Pew Research survey showed that only 26 percent of third- or higher-generation adults with Hispanic ancestry had parents who "often" encouraged them to speak Spanish. I have never been to Mexico and, aside from family reunions on Easter, Thanksgiving, and Christmas, don't often attend Mexican cultural events.

In the same respect, when I am around autistic people, I simultaneously feel a sense of ease and tension. While writing this book, I traveled to Ortonville, Michigan, for Autspace, a retreat for autistic adults. In the woods, replete with a campfire circle, basketball courts, and a mess hall, I met autistics who ran the gamut of the spectrum (this is where I met Timotheus Gordon in person for the first time, along with Charlie Garcia-Spiegel and Kris Harrison). I also met other autistic people with higher support needs, people who had limited verbal capacity or used keyboards to communicate. Some of them were professors, some still undergraduate students, others unemployed. But we all were autistic, and we all felt safe and able to be fully ourselves. But at the same time, I worried I was not steeped enough in the history of autism to fit in this space. I imagine I would react similarly if I went to Mexico; I would feel a deep connection to my roots by seeing the source of my family but also fear I was too far removed from my origins, as if I were not "authentically" Mexican.

I still struggle with these feelings, but as time progresses, I've realized there is no "proper" way to be autistic or Latino or male. I see this in my family, which is very much a symphony of identities: Stephanie's husband, Benny, is the son of an Italian American immigrant and an American-born woman, and I love spending Christmases with them since they moved to Virginia. My father's soon-to-be wife is an Asian American immigrant. Bob is the son of an Irish American immigrant and is firmly proud of his Celtic heritage. All of which is to say that my family's heritage is today on as broad a spectrum as autism exists. Our lives are a tapestry of our own lived experiences; we weave together different threads until we form an individual identity made from the disparate cloth that blends into our own insignia to tell the world who we are. It is by recognizing these various materials from which we are born that we can understand our distinct communities and how they shape our own perspectives.

9

"Till the Next Episode"

WHAT COMES NEXT

Washington, DC, is a city of suck-ups, cynics, and strivers. Social capital courses through the veins of the nation's capital, and, aside from campaign cash, personal relationships are the preferred currency in the halls of the Capitol. It is a town where people greet you, even if they are meeting you for the first time, by saying "Good to see you" —because chances are, one of you already follows the other on Twitter. The first question you are always asked is "What do you do?" but the unspoken follow-up is *How can I get ahead?* The question is a way of measuring whether you are worth their time and energy, even when it is at a party.

At times, this type of networking can be a lot to handle for an autistic person such as myself (especially as I enter my thirties). It isn't just the difficulty of navigating conversations and determining if someone wants to be my friend or if they're just trying to get something out of me; often, these parties are in cramped spaces with blaring music and lights that are either far too bright or too dark for me to make out anyone's face. Still, these gatherings are a large part of living and working in DC, especially when you are an up-and-coming journalist trying to build connections in a new and unfamiliar city.

It was at one of these riotous parties in 2015, held by two reporters for the *Daily Beast,* that I found the impetus to begin writing about autism. These parties were a veritable who's who of young Washington. I bumped into one of the hosts, Tim Mak (who would become infamous after Donald Trump's lawyer and fixer Michael Cohen threatened to "take [him] for every penny you still don't have"). He and I had met only once or twice before, and like a good host, he offered me a drink. I declined, explaining that I didn't drink because I was on the spectrum and medicine I took doesn't go well with alcohol. Instead of giving me grief about not drinking, Mak mentioned he knew tons of autistic people in journalism in Washington and suggested I write a piece on it. I was actually taken aback by how accepting Mak was. Furthermore, Mak thought autism was newsworthy enough to be written about and thought I was the right writer for that. I agreed it was a good idea but figured I wasn't smart enough at the time; maybe one day in the future.

At that point, I was still a correspondent at *National Journal,* where I was enjoying my job covering economic policy, finance, and politics. But in July, *National Journal*'s management announced the print edition would be shuttered at the end of the year. With only a few editions left, Richard Just, the print magazine's editor, made a call for audacious pieces, which led me to pitch the autism story Tim had suggested. I pitched what was supposed to be a fun, chatty piece about life in Washington, DC, as an autistic person, but Richard wanted me to go bigger. Eventually, the article evolved into a think piece in which I suggested society should stop trying to cure autism and instead focus on helping autistic people have more fulfilling lives. Richard loved the idea and told me to run with it, something that completely reoriented my trajectory as a journalist and in life.

"I Read the News Today, Oh, Boy"

At that same time in 2015, culture was changing around autism. One day, my mentor Ron Fournier, a columnist at *National Journal* whose son was on the spectrum, dropped a book on my desk called *NeuroTribes: The Legacy of Autism and the Future of Neurodiversity*, by Steve Silberman, whom I would eventually interview. Steve's interest in autism began when he wrote an article for *Wired* in 2001 called "The Geek Syndrome" that chronicled what seemed to be a boom of autistic kids in Silicon Valley. Silberman said he was moved to write the book because he noticed a gap between the lack of services for autistic children and mainstream media reports about autism, which focused mostly on the mistaken connection to vaccines.

Silberman wondered why the media was so fascinated by what caused autism while so many people suffered because of a lack of understanding about the condition. At some point, Silberman realized that part of the obsession was tied to the increase in diagnoses brought on by the changes to the *DSM* in the nineties. Even organizations like Autism Speaks used a graph shaped like a hockey stick to show a spike in autism diagnoses, which led to panicked news articles about an "epidemic."

"I thought, *Somebody has to figure out why the estimates of prevalence went up so steeply.* I just didn't see anyone doing that," Silberman told me. He began researching and stumbled across scholarly articles debunking the idea of an autism epidemic. "So, the more that I got into it, the more I had a gut feeling that something had gone wrong in autism history and that there was some information that had not been properly disseminated," Silberman said.

While writing the book, Silberman began to meet, interview, and interact with autistic people. He was invited to Autreat, the retreat in the woods for autistic adults, where he met a wide array of people on the spectrum, some of whom could not speak. The experience challenged his own preconceived notions—for instance, that autistic people didn't date. While there, he met a couple who would rock back and forth and stim together and who later married.

Despite not being autistic himself, as a gay man, Silberman felt he understood the plight of the autistic community. He too had experience with demeaning portrayals in the media, and when he was in high school, homosexuality was even listed in the *DSM.* Of course, as we saw earlier, plenty of autistic people are LGBTQ and experience a double portion of discrimination. The desire to eliminate the traits that make autistic people unique is rooted in the same impulse to suppress people from affirming their gender identity or sexuality.

It was through Silberman's time as a teaching assistant to Beat poet Allen Ginsberg that he saw what a gay man could do when he lived visibly. Today he lives with his husband, Keith Karraker, in San Francisco.

"If you think that homosexuality or autism is rare, and these people are weird, then you don't necessarily feel compelled to become part of a movement to recognize their autonomy and their civil rights," he explained. "But I knew that the [autistic community] was one of the largest minorities in America, easily." Silberman's book came out just two years after the most recent version of the *DSM* placed all autistic people under the larger diagnosis of autism spectrum disorder and helped increase the public's understanding about autism. Steve's book was not the only book about autism at the time to focus on the narratives of autistic people. Lydia X. Z. Brown and Morénike Giwa Onaiwu, collaborated with E. Ashkenazy to release *All the Weight of Our Dreams,* a collection of writings that chronicled the experiences of autistic people of color. Barry Prizant's *Uniquely Human* also emphasized a focus on celebrating autism and not seeing it as a deficit.

He explicitly wrote that "autism isn't an illness. It's a different way of being human. Children with autism aren't sick," and the best way to help them wasn't "to change them or fix them. We need to work to understand them and then change what *we* do."

Similarly, media discourse around autism has shifted. A 2020 study in *Disability and Society* surveyed 315 articles from the *Washington Post* from the beginning of 2007 to the end of 2016. The study found that over the decade of articles, there were increased mentions of neurodiversity and more inclusions of autistic people's perspectives; articles describing autism's strengths and discussing accommodations also increased, and there were fewer mentions of autism's deficits, causes, and risk factors.

The study showed that progress isn't complete—the terms *high-functioning* and *low-functioning* continue to appear despite autistic people's objections—but it does signify that the *Post,* the hometown paper for the nation's capital, is getting better at including more perspectives. Anecdotally, I can vouch for this progress because of my work at the organization. During the last year of the study, I was contacted by the paper's PostEverything section to write a piece on autism, sex, and consent that was published in April 2017. Two years later, I joined the paper full-time and witnessed how, as a news outlet, it continued to improve its coverage. When I worked as an editor, I tried to amplify autistic and otherwise disabled voices, which is particularly important at a newspaper that holds so much sway in politics.

Of course, the *Post* is just one newspaper. But the change of language there is emblematic of how media coverage has shifted. The press now sees autistic people as definitive authorities whose experiences give them expertise on the condition. This in turn can help shape how policymakers view and understand autism in the political sphere.

Campaign Season

In 2016, the political culture was changing drastically. The year after *NeuroTribes* was published, former Secretary of State Hillary Clinton became the first presidential candidate to release a full-fledged autism-policy plan. Autism wasn't new to Clinton, who spent a year working at the Yale Child Study Center when she was in law school. That was back when so-called refrigerator parents were blamed for autism, but after she married her husband and moved to Arkansas, Clinton said she began spending time with the mother of an autistic child and started to question the veracity of the refrigerator-parent theory.

In 2007, when she ran against Barack Obama for the Democratic Party nomination, Clinton adopted much of the common rhetoric about "combating" autism and putting "money into finding causes and cures of diseases from cancer to autism," even as she spoke about her various proposals to assist autistic people. This was likely reflective of the fact that the most vocal autism advocates were not autistic themselves. But her policy approach shifted drastically by 2016. By then, her campaign plan read like a wish list of issues that self-advocates were fighting for, such as a nationwide survey of autistic adults, banning chemical and physical restraints on autistic kids in schools, and promoting an Autism Works Initiative.

This shift in priorities is largely thanks to organizations like the Autistic Self Advocacy Network and the work of people like Julia Bascom, the organization's current executive director. She explained to me that she does advocacy work in a way that differs from others.

"There is a huge physical and cognitive cost for the amount of time that I spend using words," she said. Prior to holding this job, Bascom said, she spoke on average ten hours a week, but now she speaks about sixty hours. "My job has a lot of things and a lot of parts to it," she said. "What's important for people to know is that I can do those things not because my disability doesn't impact it but because I have this elaborate scaffold of supports that is invisible to people who aren't involved in it. That makes it possible."

Bascom, Ari Ne'eman, and the rest of the staff at ASAN, as well as many other autistic self-advocates, have worked for years to ensure that autistic people are not just included in the policymaking that determines their fate but at the center of it.

All of this set the stage for the role autistic advocates would play in the 2020 presidential election, particularly in the 2020 Democratic primary. Disability rights advocates gained an incredible amount of political capital among Democrats in the first year of the Trump administration after they vigorously opposed the repeal of the Affordable Care Act, Obama's signature achievement, as well as deep cuts to Medicaid. (During one of our interviews, I noticed a sign in Bascom's office from those protests that said PLEASE SAVE OUR HEALTH CARE SO WE CAN STOP MAKING PHONE CALLS.)

Many Democratic candidates consulted autistic self-advocates like Ne'eman and Bascom along with other disability rights advocates during their campaigns. Bascom said she was consulted by candidates ranging from mainstream Democrats, like Senator Cory Booker and Mayor Pete Buttigieg, to progressives, like former San Antonio mayor Julián Castro, Senator Elizabeth Warren, and self-described democratic socialist Senator Bernie Sanders.

This level of inclusion was a culmination of years of work by autistic people to be accepted as legitimate by the larger disability rights movement. In the early years of ASAN, Ne'eman said the organization focused on ending restraint and seclusion on students, not only because it was an important issue for autistic people but because it established ASAN as an advocacy organization that could represent the needs of autistic people in political discourse.

"Maybe two things that I consider to be the greatest achievements

I had over my time running ASAN—the first would be building the political, financial, and nonprofit infrastructure to take our community into the mainstream instead of the fringe," he said. "And then the second would be wedding and intertwining autism into the disability rights movement."

I've covered politics long enough to know most campaign platforms don't mean candidates can actually implement them; many require legislative action or working with opposing parties or even factions within the same party. Rather, they are aspirational.

"I think that people need to remember that primaries are about articulating a vision and setting the bar as high as possible and getting people to compete with each other," Bascom said.

But despite the combination of a competitive primary and disability rights advocates' increased sway, the eventual Democratic nominee was former Vice President Joe Biden. Trump had won traditionally Democratic states like Wisconsin, Pennsylvania, and Michigan in 2016, and many Democrats thought Biden could win back those states.

Biden's presumed frontrunner status made it less necessary for him to court interest groups, since Democratic voters' primary concern was defeating Trump, not advancing their preferred policies. Biden did not release a comprehensive disability plan during the Democratic primary beyond a webpage full of general principles on his campaign website (this despite the fact that he voted for the Americans with Disabilities Act when he was a senator in the 1990s and has spoken movingly about his own disability, his stutter). However, Bascom told me she spoke with Biden's team during the primary. Still, Biden was criticized during the primary for ignoring disability rights activists, prompting the hashtag #AccessToJoe on Twitter. Ironically, despite Biden's frequent invocations of his time as Barack Obama's vice president, many of the most vocal advocates, like Ne'eman—and other disability rights advocates, like Rebecca Cokley and Maria Town—were alumni of the Obama administration.

Biden finally released a disability policy platform in June 2020, long after his biggest opponent, Bernie Sanders, had dropped out of the race. Like many of Biden's other policies, his plans were not as aspirational as Warren's, Sanders's, or Castro's. But his approach did support phasing out subminimum-wage labor and asking the U.S. Justice Department to review guardianship laws and promote alternatives like supported decision-making. Biden also released an entirely separate platform specifically focused on people with disabilities during the coronavirus pandemic.

But at the same time, when Biden—who has been known for making verbal gaffes that range from the benign to the outright nonsensical—was rolling out his disability plan, he said that "everybody has some form of disability," which is patently not true. While it is true that disability affects everyone's lives (disability rights advocates often say that it's one of the few minority groups you can join), saying everyone has some form of disability flattens the experiences of people whose disabilities prevent them from living their lives. If everyone has a disability, then disabled people have no reason to complain.

But once Biden won the Democratic nomination, his campaign made overtures to the disability community. In July of the campaign, he released a plan that would eliminate the waitlist for home- and community-based services through increasing money for Medicaid and supported creating a "innovation fund" to find alternatives to institutional care. For the most part, though, Biden's campaign (somewhat understandably) focused on other topics, such as the coronavirus pandemic and protests against racial injustice. But in both cases, disabled people faced the brunt of the pandemic. NPR reported at the time that people with intellectual disabilities and autism who contracted coronavirus died at a higher rate than the rest of the population in New York and Pennsylvania.

Biden also hired Molly Doris-Pierce, who was disability outreach coordinator for Elizabeth Warren and whose disability plan was widely praised by the disability community. Doris-Pierce worked closely with Ne'eman and Bascom when Warren crafted the plan, thus validating to the self-advocacy movement. Doris-Pierce, who has obsessive-compulsive disorder, said working with the disability community has helped her accept the way her brain is wired.

"I didn't want to claim that title, but through working with activists, the cycle has been incredibly helpful for me to be like, 'Oh no, this is considered a disability,' and it does shape the perspective about these issues," she told me. Through her work in politics, she now identifies as disabled. "It just never occurred to me until I was an adult, naively, that there are abled people who don't have any experience in that realm." Doris-Pierce's experience shows how disability rights not only liberates people who are openly disabled but also allows those who may not realize they are disabled to accept themselves and their identities and eventually try to expand accessibility to others.

Doris-Pierce hosted a digital watch party for disabled supporters during the second night of the Democratic National Convention. Interestingly, two of the surrogates who attended the one watch party I went to were Representative Ayanna Pressley of Massachusetts, who supported Warren's campaign and twice said "Nothing about us without us," the slogan of the self-advocacy movement for disability rights, and Castro, who had a contentious exchange with Biden during one primary debate. But neither spoke at the main convention.

On the other side of the aisle, the Trump campaign did little outreach to the disabled community. At an American Association of People with Disabilities summit, Trump's daughter-in-law Lara Lea Trump did highlight that under his administration, the Food and Drug Administration banned shock therapy. Lara said it was a "true shame that it took until now to ban this cruel practice."

Biden's selection of Kamala D. Harris as his vice president was particularly significant because she was the first Democratic presidential candidate to release a disability policy, one that was arguably more ambitious than Biden's. But Harris also received criticism from the disability community toward the end of her own doomed presidential campaign when she released a mental-health reform plan that called for the repeal of the "Institutions for Mental Disease Exclusion." The IMD exclusion prevents Medicaid from funding inpatient care at facilities with more than sixteen beds. The goal is to prevent mentally ill people or those with psychiatric disabilities from being crammed into large institutions or excluded from the community. When Harris included this in her plan, disability rights activists worried that repealing the exclusion could lead to the return to institutionalization that they had long opposed. Harris was planning on reworking her mental-health plan after the intense criticism, but she ended her campaign before it could be released.

Despite the political climate, though, there are reasons for autistic people to be optimistic. As an article in the *Atlantic* that focused on Biden's stutter read: "This evolution in treatment has been accompanied by a new movement to destigmatize the disorder, similar to the drive to view autism through a lens of 'neurodiversity' rather than as a pathology."

Tim Alberta, a *Politico* reporter (and a former colleague of mine from *National Journal*) asked questions about disability during a December 2019 Democratic primary debate while wearing a pin that supports the Autism Alliance in Michigan, his home state. Other reporters, like Abby Abrams at *Time* and Maggie Astor at the *New York*

Times, rigorously covered disability throughout the 2020 campaign. These well-regarded journalists are equipped to cover disability politics and hold those in power accountable however it changes.

For my own part, while an editor at the *Washington Post,* I had the capacity and wherewithal to know whom to call whenever autism comes up in the news cycle. For example, during the coronavirus pandemic of 2020, I asked David M. Perry, a disabled father with an autistic son who also has Down syndrome, to write about the challenges of procuring enough of his son's favorite foods and supplies to last during the quarantine. I wasn't taking a side—that's never my job as a journalist. Rather, I made sure to amplify the voices of people who are legitimate participants in America's democratic process.

Having disabled and autistic writers and editors means that news outlets can flag items that would normally get ignored or receive minimal coverage. In 2019, I was working a normal shift as a Saturday-evening editor at the *Hill* when I saw a tweet from right-wing provocateur Ryan Saavedra about Senator Kamala D. Harris. At a campaign event in New Hampshire, a man asked her about President Donald Trump and what she was going to do to "diminish the mentally retarded action of this guy." Harris responded: "Well said."

"I hope @KamalaHarris enjoys losing the disability vote," Bascom tweeted, then added, simply, "Shameful."

My colleague Zack Budryk, who is also autistic, happened to be working that Saturday as well, and I knew he would be up to handling this story correctly. I knew Budryk was right for this story not just because he is a good reporter but because of his lived experience. Neither Budryk nor I are advocates, and we are not partisan in our work; we are journalists, and it is our job to portray the world as it is. But being autistic and steeped in our community allows us to consider topics that might otherwise be ignored. We know whom to call, whom to consult, and who is a credible authority; it's similar to how people from diverse backgrounds help journalists tell the stories of those communities more accurately.

The protests for racial justice in 2020 sparked many discussions about the lack of racial diversity in newsrooms across the country, not just in terms of staffing but also in leadership (a concern I took very seriously as an assistant editor and a Latino at a highly respected newspaper). In the same respect, the swath of disability-related stories that are regularly ignored or get covered from abled-person's point of view underscores the disabled voices in newsrooms.

Being in a newsroom with another person on the spectrum was comforting since it made me feel like I had my finger on the pulse of something. When I first came to DC, there were very few other openly autistic journalists (I am positive there are plenty of autistic and otherwise neurodivergent people in the news business and in politics writ large, but it isn't my business to ask or probe or prod them about it). Often in the past, I found writing about autism difficult because much of it required firsthand reporting or on-the-record statements, and it meant showing readers how many claims, rhetoric, and language about autism were simply not true.

Even when my first big piece about autism was published to much fanfare, I still worried about being one of the only autistic reporters, because if I messed something up, it could misinform so many people. I was burdened by the fear of letting my community down.

Those fears were unfounded because, as it turns out, I am not alone in this movement. Thanks to other autistic reporters, like Budryk, Dylan Matthews at Vox.com, S. E. Smith, and Sara Luterman, I am not solely responsible for covering autism. I have not been pigeonholed by my bosses either, so while I occasionally cover disability, I still mostly write and edit pieces about politics. Nowadays, more major news outlets quote autistic self-advocates like Ari Ne'eman and Julia Bascom and cover parents like Shannon Des Roches Rosa who are not searching for a cure for their autistic children but rather trying to help them live fulfilling lives. Similarly, journalists across the country often reach out to me when they're assigned a story on autism, and I gladly connect them with the right resources so they can write about the topic accurately.

Parent Advocacy

The increased visibility of autistic people in public life has empowered parents to have new role models of how to raise autistic children. One such parent is Roberta "Robbie" Kaplan, who is probably best known for arguing *U.S. v. Windsor* in front of the Supreme Court, which resulted in the Court striking down the federal government's ban on recognizing same-sex marriages as unconstitutional.

I was interested in speaking to Kaplan because, while her legal work was focused on ensuring the rights of one disenfranchised group—LGBTQ people—she is also the mother of someone from another marginalized group. When Kaplan found out her son was on the spectrum, she compared it to coming out—both experiences were clouded by fear and stigma. "The only way to eliminate the stigma and the shame is for more and more people who are on the spectrum to come out of the closet, for lack of a better phrase," Kaplan told me.

As Kaplan and her wife, Rachel Lavine, began to learn about their son's condition, they noticed how many professionals suggested that their son be exposed to things that bother him and just "grin and bear it." Kaplan, who suffers from migraines, compared it to going to a smelly fishing pier and being expected to power through.

"It's extraordinary to me that today, even in Manhattan among the educated wealthy classes, parents are still getting that kind of advice," she said. "Rachel and I have consciously decided to listen to people on the spectrum and adults on the spectrum and take what they're saying as far more important."

Hearing Kaplan, an advocate for LGBTQ people, say that she and her wife chose to listen to autistic people was refreshing. It showed that she considers autistic people the experts about their condition and the ones who can help her be a better parent to her son. It also means she accepts her son's autism not as something to be fixed but as an essential part of his nature. She wants to help him live the best autistic life possible rather than a life measured by nonautistic standards.

Kaplan and Lavine's ability to accept their son is a drastic change even from how things were when I was growing up. Back then, while my mom and countless other parents navigated the new diagnostic criteria, there were few openly autistic people for parents to turn to for advice. While my mom never resorted to any harsh or draconian techniques like secretin, and I can't defend parents who put their children through horrific treatments, I understand how it is possible for parents to pay heed to false prophets. There are nonautistic people who want to reap profits from terrified parents of autistic children who think an autism diagnosis is a worst-case scenario.

The best response, then, is to have public examples of autistic people who occupy the entire spectrum and who require various support needs. We also need more examples of families who accept their autistic relatives for who they are. A mother like Shannon des Roches Rosa, who loves and accepts her son Leo, is a perfect example of a parent who once was afraid of autism but chose love instead. People like Hari Srinivasan can show parents that their children are not "trapped" inside an autistic mind but that they just need to find a new way to communicate with them. Having positive examples like the ones in this book can show parents and families not only how to manage the difficult parts of autism but also what is possible when they support their loved ones.

"Both Rachel and I grew up as people who really have spoken up for ourselves and who believe in speaking up for ourselves," Kaplan added. She and Rachel want their son to be that way too.

Similarly, toward the end of my interview with Ari Ne'eman, he asked how old I was when I was diagnosed. At the time, I was on vacation with my mom and sister in Jacksonville, Florida, and had been talking about the diagnostic process with my mom.

"I was diagnosed at eight—at seven or eight, yeah," I told him.

"So, I was diagnosed at twelve," he said. "And what's interesting to me is that, if you go to folks who are maybe five or ten years younger than us, a lot of them were diagnosed even earlier than both of us, right?"

"Yeah, yeah," I said.

"And so they've had many of the same experiences that we've had and also some other ones, right?" he said. Ne'eman pointed out that the kids from the generation after ours—kids like Robbie Kaplan's son—likely had many of the same experiences that we did, like dealing with backward school systems where many autistic kids are still restrained and secluded. Many of them (like Eryn Star from chapter 7) grew up with guilty parents who believed their children were autistic because of vaccines. The generation after ours grew up with clearer diagnostic criteria for autism spectrum disorder and were born after crucial legislation like the Individuals with Disabilities Education Act was passed. Now, there are more spaces for autistic people, whether that's organizations like ASAN or groups on college campuses.

"I remember, my first semester in college, I went to the local Hillel, and I knew this was the place I could go and connect with other people who were Jews like me," Ne'eman said. "But there was no feeling that anywhere, let alone on my college campus, there was anything comparable to that for autistic people."

In the same way, when I went off to UNC, I didn't see many autistic-led organizations. When I briefly talked with some kids from a campus autism group, I realized there were no other openly autistic people in the group itself; it was seen as more of a charity activity. I never went to a meeting, so it's true, there were not many places to be accepted even then. However, kids from the generation after mine grew up understanding their autism and eventually owning it. There are kids, like Greta Thunberg, who are aware that their autism makes them better advocates. And kids like Chris and Cori Williams' children, who grow up with families that love and accept them, and Lydia Wayman, who has people who support her.

"That is absolutely amazing to me," Ne'eman said. "They've grown up with this, with being connected to a community that has agency in a way that I never had."

I've noticed this new wave of openness in my own life too. One of the most magnificent things that has happened because of my writing is that now, many college-age journalists reach out to me to ask about breaking into the business as an autistic reporter. It is a model that didn't really exist for me when I was in journalism school, not because there were no autistic journalists—certainly there were—but there were few who were open about it and who wrote about it regularly. I never wanted to write about my autism to promote myself, but if I can give some inspiration to would-be autistic reporters, then I know I have done a service, which is why I wanted to become a journalist in the first place.

I still regularly get e-mails and messages from parents of autistic kids who want advice or are just grateful that I offer an example of a different way to be autistic in this world. I'll carry these messages with me forever. I hope that with this book, I've shown that there is no correct way to be autistic and that autistic people who have higher support needs or those who cannot speak or hold full employment deserve as much dignity and respect as I am afforded.

And the story for Generation Z and beyond will be unique. To be certain, many of them may live with continued challenges, like the persistent misinformation about vaccines or incomplete portrayals of autistic people in pop culture. But they will have grown up with concepts like neurodiversity as part of the lexicon. Hopefully, they will not have to wait until their twenties to accept their autism (like I did). They will live in a world with fewer institutions and cruel treatments. Through the work that their peers and predecessors have done, they will have role models setting an example for how to be autistic in a world that is not built for us. I already see this new generation's work on the internet, and I'm so impressed by their willingness to embrace the parts of themselves that sent their ancestors to institutions. They will blaze their own path far beyond any horizon I can see.

I hope they find ways to bend the world to fit them. As the conversation continues to evolve, I hope new voices will join in and add to it. My earliest writings about autism feature some concepts that I now find unsavory (like talking about a cure or functioning levels). Meeting autistic people in my journeys as a journalist changed how I view autism and, in turn, myself. My hope is that I can do the same for the next generation of autistic people while giving them a working language to understand themselves. At the same time, I trust that they will refine the grammar in a way they best see fit.

At the close of the Civil War in 1865, Frederick Douglass, the runaway slave turned abolitionist, addressed the Massachusetts Anti-Slavery Society, where he vociferously called for giving Black people the right to vote. "What I ask for the Negro is not benevolence, not pity, not sympathy, but simply justice," Douglass said.

It would be daft to compare autistic people's tribulations to those of enslaved people and the experiences of Black Americans (not to mention the many black autistic people who face myriad difficulties balancing all of their identities). But I do see a similarity in how autistic people have been historically dehumanized. I think back to Jessica Benham, who overheard a donor say that people in institutions were not "fully human," and Ole Ivar Lovaas, the psychologist who helped pioneer the use of applied behavioral analysis in children, who said that autistic people "are not people in the psychological sense."

For years, this toxic way of thinking has denied autistic people the chance to speak for themselves and articulate what their desires are. Often, if they are like me, society tells us we are not "autistic enough" to understand the needs of autistic people who require more support. However, if they require more support or cannot speak, then their agency is denied altogether. Either way, our attempts at self-advocacy are rendered illegitimate, and parents are given the microphone instead.

Douglass concluded his passionate oration on voting rights by asking, "What shall we do with the Negro?" to which he had a simple response: "Do nothing with us."

Here now, I'd like to make this same request for the autistic community. People who are not autistic often assume they are acting benevolently by hand-holding those on the spectrum. But despite their best intentions, there is an element of condescension in these actions because it assumes that nonautistic people know what's best. But it is autistic people who live with the condition of autism—for all of its positives and negatives—as well as with the consequences of any collective action meant to help them. If there is going to be policy that has seismic impact on their lives, they deserve to have a say in it, no matter how they communicate. Furthermore, while many parent advocates, clinicians, and other "experts" may have good intentions, centering their voices continues to give them power that should lie with the autistic community. To achieve any true sense of freedom, autistic people need to take this power back.

This book does not seek to advocate for one policy proposal versus another; policies that affect autism are multifaceted and should be debated in the public sphere. But what should happen is that when autistic people articulate their needs (however that articulation is done), they should be listened to as legitimate voices.

Epilogue

In my last month of writing this manuscript, two disparate things happened. The first was that Mel Baggs died. Mel (whom I wrote about in chapter 1) was an advocate and the first autistic person I ever saw on the news. When I heard about their passing, I remember feeling overwhelming sadness. Though I never met them, I grieved knowing that they died without knowing what they meant to me.

I knew Mel got a lot of press for a YouTube video they uploaded in 2007, but I'd never watched it in its entirety. Now, with the news of their passing, I pulled up the video. The first half features Baggs making noises with their voice while also stimming in multiple ways —running their fingers across a keyboard, scraping their fingernails against the steps of a stepladder, nuzzling their face into a magazine, and many other mannerisms.

These were all things I immediately recognized. Most people don't understand how soothing a magazine page against the face can be or how much the tension and friction of a book page can excite the senses. Many people don't get how running your finger under cold water can make you feel completely alive.

The second part of the video features a voice narrating Baggs's typed words as they run their hands through the water, experiencing sensory processing like seeing, tasting, smelling, feeling, and hearing. Baggs's point was to show that there are multiple forms of communication.

"Ironically, the way that I move when responding to everything around me is described as 'being in a world of my own,'" they say in the video. "Whereas if I interact with a much more limited set of responses and only react to a much more limited part of my surroundings, people claim that I am 'opening up to true interaction with the world.'"

Similarly, Baggs noted the hypocrisy that people like them were taken seriously only if they communicated in the language neurotypical people understood. "I find it very interesting that failure to learn your language is seen as a deficit but failure to learn my language is seen as so natural that people like me are officially described as mysterious and puzzling rather than anyone admitting that it is themselves who are confused."

Baggs was insistent that this video was not supposed to be a "voyeuristic freak show" demonstrating the "bizarre workings of an autistic mind."

"It is meant as a strong statement on the existence and value of many different kinds of thinking and interaction in a world," they said. This brief statement, and Baggs's decision to show the way they communicated without words, articulated everything I wanted to say more eloquently than I ever could. Baggs's existence lit a spark in me that swelled into a wildfire that is my life's work.

The other thing that happened was that Drew Savicki—the autistic political prognosticator whom I profiled in chapter 3—finally landed a job after years of searching. Throughout the 2020 presidential election, he contributed articles to 270toWin, a website most famous for having a map of all fifty states that allows people to play around with hypothetical votes to see different ways to win the electoral college (I use this site to game out multiple scenarios for elections). Savicki's pieces offered comprehensive breakdowns of states, their political trajectories, and why they vote the way they do. In addition, he occasionally does side work for Democratic elected officials.

Despite not having a college degree, Drew has carved out a niche for himself as a brilliant and respected election analyst, and I consider him one of the best out there because he does not let his own biases get in the way of his attempts to game out races. His role at 270toWin.com ended after the election but it cemented his status in politics. I want him to succeed and want him to be recognized for the skill I know he has. I'm thrilled for Drew, especially since I've known him for some time and understand how frustrating it can be to have talents and skills but not find a job with which they connect.

Savicki, Baggs, and I are supposed to have different variations of autism; I am supposed to be considered the "highest functioning," Savicki is considered somewhere in between, and Baggs was supposedly "low-functioning." But seeing how much Baggs's stims matched mine and knowing how Savicki's own skill in politics surpasses mine, I realized how little difference exists between us.

We all require support, both in terms of services and social support from those around us. All three of us—and all autistic people, for that matter—require people treating our communication and our talents as legitimate. What makes the difference, ultimately, is if people consider how we articulate our needs as valid.

"There are people being tortured, people dying because they are considered non-persons because their kind of thought is so unusual as to not be considered thought at all," Baggs said at the end of their video. "Only when the many shapes of personhood are recognized will justice and human rights be possible."

Afterword

A few months before the coronavirus pandemic shut down the world, Greta Thunberg, the Swedish climate-change activist who is autistic, sailed across the Atlantic to give her righteously angry address at the United Nations. Before she went to New York, Thunberg visited my adopted hometown of Washington, DC, for a march around the White House Ellipse. Since I was a night editor for the *Hill* at the time, I decided to pay a visit to perhaps the most famous autistic person in the world. I and other reporters tried to speak with her but were ultimately unable to.

At the time, I spoke to another journalist who also happened to be autistic and we both marveled at how Thunberg was able to participate in these marches without headphones or anything to mitigate the noise. It was affirming to see an autistic teenage girl lead such a well-organized protest for action against one of the greatest challenges facing humanity. Thunberg has a sense of self and her place in the world as an autistic person that I lacked at her age. The way she sees her autism as a catalyst for her activism shows how a world that views autistic people as fully human is also a world where they can ignite action.

At that same protest, an elderly woman stood nearby wearing a sign that said, WAR CAUSES CLIMATE CHANGE but below said, VACCINES DANGEROUS: LEARN THE RISK 1–34 AUTISTIC CHILDREN. I later asked her about the sign and she rattled off the usual anti-vaccine talking points: autism used to occur in only 1 in 10,000 kids; there were not as many vaccines available when she growing up; better quality of living, not immunization, has reduced disease.

Of course, as we discussed throughout this book, none of this is true: increased diagnoses were a product of broader diagnostic criteria, better diagnoses, and better services being offered in school, as well as the narrowing of the racial diagnosis gap. But this dichotomy stuck with me: Here was a woman spouting absolute nonsense about autism that had repeatedly been debunked—showing how deeply entrenched myths, misperceptions, and malicious ideas about autism persist despite new information proving them wrong—and, ironically, she was standing a few feet away from a young woman who frequently preaches that her autism is an asset for her.

I finished my research and turned in the manuscript for this book just as the COVID-19 pandemic began to rage across the globe, which is why I barely touched upon it in the initial edition. But as COVID ravaged communities, I thought about that demonstration in DC and about the content of this book for multiple reasons. For one, autistic people are far more vulnerable and likely to die from COVID-19 than neurotypical people, especially those in congregate care facilities. Had there been sufficient money for home- and community-based services and no waitlists, it is entirely possible that more autistic people (and other disabled people) would be alive today, and it will likely take years to get a close approximation of how many of them perished. These unnecessary deaths show how utterly disposable autistic people's lives are, and thinking about them made my book feel all the more urgent.

In response to this incredible loss, President Joe Biden proposed spending a whopping $400 billion on home- and community-based services as part of his signature infrastructure proposal. However, the bill was eventually split in half as a means of convincing Republicans to pass a "hard" infrastructure bill with bipartisan support. Biden would sign this the traditional infrastructure bill in November 2021. Meanwhile, the second proposed bill slashed funding for numerous provisions, including HCBS funding, which was reduced to $150 billion, all to placate more moderate members of his caucus—only to have Senator Joe Manchin of West Virginia kill the legislation in December 2021 (this despite his state's history of promoting the well-being of autistic people, as I chronicled in this book).

At the same time, many of the same bad actors who propagated the myths about the link between vaccines and autism were also spreading myths about coronavirus and the vaccines developed to combat it. Indeed, the hellscape of disinformation that seeks to erode our trust in everything from vaccines and medical information to the sanctity of our elections is the fruit from a tree that was planted in the soil of the autism vaccine panic. Neurotypical society is now reaping the harvest that anti-vaxxers sowed. I began promoting the book in August 2021, when lies were still being spread about COVID-19 vaccines, and I pointed out that the false narrative about autism and vaccines that dominated the 1990s and 2000s was what *The Hobbit* was to *The Lord of the Rings* (with fewer Orcs). Ironically, anti-vaxxers' lies about vaccines and autism made me all the more enthusiastic about receiving my single dose of the Johnson & Johnson vaccine and then my booster shot of Pfizer.

Andrew Wakefield, the discredited doctor who wrote the original

Lancet study before losing his medical license, spread misinformation about COVID-19 vaccines on a QAnon Twitch stream, urging people to take drugs like hydroxychloroquine instead of receiving a vaccine. Robert F. Kennedy Jr., the former environmental lawyer turned anti-vaccine activist, similarly helped to boost anti-vaccine myths, including with the odious claim that baseball legend Henry Aaron's death was linked to the COVID-19 vaccine. (There is, of course, an irony to Robert Kennedy Jr.'s opposition to vaccines, given that his father helped to shed light on abuses at Willowbrook State School in New York, and his uncle, Senator Edward Kennedy, was an architect of the Americans with Disabilities Act.)

Meanwhile, Del Bigtree, the former television producer who produced *Vaxxed,* a documentary lionizing Wakefield, spoke at a small rally on January 6, 2021, in Washington that coincided with the insurrection at the Capitol, where supporters of President Donald Trump attempted to overturn the election. In his speech, Bigtree linked myths about the 2020 presidential election being stolen with misinformation about COVID-19. It is, to borrow from Lincoln, altogether fitting and proper that Trump's presidential ambitions would begin with him spreading malarkey about the link between autism and vaccines and would be bookended by vaccine fearmongers spreading the Big Lie about the 2020 election. Trumpism created a singularity that united right-wing paranoia about multiracial democracy, obsequious toady social climbers who would say anything to move up the rungs in Washington, conspiracy theorists who grifted off the general public, and pernicious liars about vaccines.

As if to accentuate how the misperceptions and outright lies around autism continue to permeate our world, when the "Stop the Steal" rally gave way to the insurrection at the Capitol, Jacob Chansley gained notoriety as the "QAnon Shaman" for going shirtless and wearing a fur cap with horns. But as Chansley's trial unfolded, his lawyer Albert

Watkins tried to use the fact that he is on the autism spectrum as an excuse for his client's actions, specifically saying, "These are people with brain damage, they're fucking retarded, they're on the goddamn spectrum."

Reading those words brought me back to Thunberg's rally. This teenage girl certainly has been maligned and demeaned; people have said she is "brainwashed" or a "marketing gimmick," as Donald Trump Jr. called her. Her anger has been weaponized against her, largely rooted in the fact that, as this book chronicled, society doesn't really know how to handle autistic girls, combined with the fact that women's anger—especially when used as a tool for political organization—is often seen as a threat. While Thunberg is ripe for targeting by many bad-faith critics, it is notable that her school strikes for climate and her rally in front of the White House never devolved into the chaos that characterized the Capitol riot. And yet Chansley's lawyer had the gall to use autism as a justification for his client's actions. Thankfully, the court saw it differently and Chansley was sentenced to prison, but the entire episode showed how the legal and political systems consistently misunderstand autism and spread stigma about both autism and autistic people.

At the core of both of those arguments is the idea that autistic people cannot participate in public or political life: if they are too eloquent or too competent in their activism, it must be because they are vessels for ulterior forces far more intelligent than they; if they engage in violent behavior, then it's because their poor little autistic brains were manipulated, and therefore, they should be granted mercy. Either way, the underlying argument is that politics is too complicated for these pitiful autistic people (the condescension in this thinking echoes those who said that women should not be allowed to vote). The truth is that Thunberg—and all autistic people—deserve to participate in political action, and people like Chansley deserve to face the consequences of their actions just like the rest of the would-be insurrectionists.

Around the time I published this book, my friend David M. Perry asked me in an interview how autistic people won the culture war. The question gave me pause because in some ways, autistic people still haven't. At the same time, they now have far more sway than they did even when I started writing about autism in 2015. This was nowhere more apparent than when Australian singer Sia released her movie *Music,* which was about an autistic girl and her neurotypical sister. Aside from the fact the film was little more than nonsensical drivel with musical interludes that did little to nothing to move the plot along, it also showed how Sia, her producers, and, by proxy, most of Hollywood utterly misunderstand autism. Sia described the movie as "*Rain Man,* the musical, but with girls," without realizing how the Dustin Hoffman and Tom Cruise movie has aged in the minds of autistic people. Nor did she cast an autistic actress to play the titular character Music in the film, but rather picked her regular collaborator Maddie Ziegler. In response to criticism, Sia tweeted that she had tried casting "a beautiful young girl nonverbal on the spectrum," but that the girl found it "unpleasant and stressful," as if autistic people's lack of capacity rather than Sia's inability to accommodate autistic people or adjust her vision according to autistic people's needs were the reason for this miscast.

What's worse, Sia was borderline Trumpesque in her responses to autistic people who pushed back, saying to one aspiring autistic actor, "Maybe you're just a bad actor." In addition, she said, "Fuckity fuck why don't you watch my film before you judge it?" The ultimate death knell for the film was the fact that it included multiple scenes of both Kate Hudson and Leslie Odom Jr. using a form of restraint on

Ziegler's character that has proven lethal for plenty of autistic children when used to stop meltdowns. In response, Sia said she would remove the scenes, and she deactivated her Twitter account.

The entire debacle signified the tectonic shift that has occurred in how people talk about autism in advocacy spaces. While the conversation in politics has shifted in recent years, Hollywood has been slower to catch on. Had anyone in the entertainment industry at any point consulted with autistic people, as Erica Milsom did when she directed *Loop,* it would have been clear how one-dimensional and harmful this portrayal of an autistic character was, and it would have been obvious that autistic people themselves would not accept being portrayed in such a way. Sia said that her movie was a "love letter," but it reeked of the infatuation that fifth-grade boys have of the cute girl in class (and I say this as a retired fifth-grade boy with a crush on the cute girl in class); Sia loved the idea of loving us without ever bothering to learn about us, or about what we needed or what we wanted. The portrayal was done without our consent, and therefore it came as no surprise that Sia's love went unrequited.

What are we to make of these times? What are we to make of the days when so many autistic people can now finally have a say in policy-making about our lives, but at the same time, far too many of us have perished in a pandemic at a devastating clip? How do we square our entrance into the public sphere in political campaigns—even campaigning on subjects that are not specific to disability—when our work is still diminished because of our condition? How do we square the crushingly patronizing way people see us on a screen with the fact that we can now speak out against them?

I regret to say I have no clear answers. But what I can say is that we as autistic people are on the cusp of our next chapter in history, and moreover, we now have the tools to write that history ourselves without an interpreter or someone else writing it for us. It will be in our words, our prose, and our sentences. We will dictate, write, rewrite, and edit this script until the next generation comes and builds upon it and corrects our errors. We will allow ourselves to be human and not let our flaws speak for the rest of our community, and we will lift each other up to the most commanding heights. We will have our disagreements, as any community as heterogeneous as ours is bound to have. But we will let our voices be heard, whether with vocal cords, with keypads, or with stimming. But we will be heard.

In my time since publishing this book, I am honored to have heard from so many autistic people and their loved ones, and I have taken your feedback to heart. I agreed with most of the criticisms autistic people leveled and have tried to listen with an open mind and heart. I hope I have a young enough mind to absorb such feedback and enough maturity not to take it personally and adapt thusly. Writing about your own community and learning about it at the same time is an inherently perilous venture, but I am grateful that this community in which I have simultaneously chronicled and immersed myself has accepted me and let me amplify its story. May the words that I have recorded resonate with future autistic generations and may they create new melodies and rhythms we never could fathom.

—Eric Garcia, March 2022

Acknowledgments

I am first off eternally grateful for my family. Throughout this book, you read about my mom's love and dedication toward giving me a better life. I can assure you that she is even better in real life and I thank God for her. If parents want to know what they can do to best advocate for their kids, they can't go wrong by looking at Debra Garcia. This book is the fruits of her labor, from teaching me how to read to ensuring I got adequate services in school and pushing me to do my best in class. In the same respect, my earliest memories of journalism are of watching the news with my father, Charlie. His regular watching of talk shows exposed me to a wider world, and the people on television looked like they were having a lot of fun. I am thankful for him and that he has found love in his life again with Helen. In the same respect, my stepdad, Bob, has constantly supported me in more ways than I can count. When he saw I found something as compelling as journalism, he sought to foster it by buying me as many books and magazines as possible. His love and advice have guided me constantly. My sister, Stephanie, has been my rock and my best friend since she was born eleven months after I was. To see her grow up into such an amazing daughter, wife, and teacher is one of the greatest joys of my life. I am also grateful to her husband, Benny Oliveri, and his family for their support during this time and all times. Hopefully finishing this book and the end of a pandemic means I can travel to see you more. Much thanks also to my extended family of cousins, aunts, and uncles through blood and marriage. So many times, when I fell down, your love lifted me back up. I lost my grandmother Connie Hernandez when I began writing the proposal for this book, and I think about her every day and I know I am alive and thriving because of her prayers and love from beyond. Meanwhile, I know my late grandmother Garcia, who died when I was nine but constantly watched the news and was politically engaged (my parents' greatest nightmare was that she would teach my sister and me about the Clinton impeachment scandal when we were in grade school), would have loved that I got into political journalism. I love and miss you both.

The second group of people to whom I owe the greatest amount of thanks is every person I interviewed, specifically all of the autistic people. Autistic people's capacity to control their own lives is constantly questioned and it often starts with their capacity to speak. So, for that I am grateful to Eryn Star, Jessica Benham, Lydia and Brisky Wayman, Sara Gardner, Andrew Savicki, Shannon and Leo Rosa, Claire Barnett, Chris and Cori Williams along with their children, endever star, Kris Harrison, Dave Caudel, Ashia Ray, Aria, Ari Ne'eman, Julia Bascom, Samantha Crane, John Marble, Amy Gravino, Arianne Garcia, A. J. Link, Morénike Giwa Onaiwu, Liane Holiday Willey, Claudia Ng, Charlie Garcia-Spiegel, Sara Luterman, Finn Gardiner, Timotheus Gordon, John Elder Robison, Lindsey Nebeker, Dave Hamrick, Lydia X. Z. Brown, Hari Srinivasan, Richie Combs, Maxfield Sparrow, Zachariah Lewis, M. Remi Yergeau, Marcelle Ciampi, Anlor Davin, Greg Yates, and Cal Montgomery. Many thanks also to nonautistic people who were willing to be interviewed or who assisted with letting me read their studies including Katharine Zuckerman, David Mandell, Doug Leslie, Molly Doris-Pierce, Tatja Hirvikoski, Tara Cunningham, Alan Kriss, Arvind Venkat, the team at the Department of Health and Human Services, Robbie Kaplan, Karen Corby, Colton Callahan, Erica Milsom, Catherine Lhamon, Anna Krieger, C. J. Ereneta, Claudia Ng, Michael Kendrick, Julie Christensen, and so many others. Much thanks also to the staff and leadership at Square, the Love and Autism Conference, Autspace, Marshall University, the Democratic National Committee, Stanford University, and Vanderbilt University for facilitating my reporting when I visited your facilities either virtually or in person. There were also plenty of autistic and nonautistic people whom I interviewed who did not make the final draft to whom I am indebted. Your omission is in no way a reflection on the importance of what you do.

In the time since, I have befriended many autistic people who were not listed in this roundup of people. Primary among them are Haley Moss, Steve Lieberman, Dylan Matthews, Andrew Solender, Zack Budryk, Noor Pervez, Cassandra Nelson, Carrie Hall, Judy Endow, and so many others. Knowing you taught me more about myself and to be kinder to myself, which hopefully allowed me to be better as a writer.

The physical act of writing a book is an often-solitary venture filled with late nights by oneself. But the act of finishing a book cannot be done without a robust team and I had the best in the business. Heather Jackson was the person who saw the potential for this book and suggested I write a proposal. She whipped me into shape as she taught me about a completely different style of writing and helped me with negotiations but was also delicate where need be. I will always owe her for believing in this project. When we began to shop for publishers, the first person we met was Deb Brody from Houghton Mifflin Harcourt and I was immediately impressed and thought it would be great to work with them. Thankfully we did and Deb was a great partner and editor. Deb handed it off to Emma Peters in the last year of this project and it was the perfect match. Emma was a meticulous editor who massaged my sentences and yielded the best possible drafts from me for each chapter. I could not have had a better team. Let's do this again sometime, yes?

This book started with an idea that began with Tim Mak, who suggested I write a piece about autistic people in the Beltway and kept insisting I write it. Tim, this one's for you. But the real adventure began when I proposed the original magazine piece to Richard Just at *National Journal,* who suggested it go from a chatty piece about DC insiders to an ambitious piece about why focusing on curing autistic people was misguided. Richard handled this with the utmost care and is now a dear friend. Much thanks also to my regular editors Matt Berman and Marina Koren, who let me take time off to write that piece as the Republican primary was taking off in 2015, back when we thought Donald Trump was little more than a joke candidate who would eventually crash and burn. The rest of the *NJ* magazine team of Molly Mirhashem, Amanda Cormier, Andie Coller (who came up with the title "I'm Not Broken," which Heather later adapted for the book title), Sarah Smith, and Sacha Zimmerman turned it into something I never envisioned it could be. The entire *National Journal* team, from mentors like Bob Moser, Ben Pershing, Kristin Roberts, Tim Grieve, and Patrick Reis to colleagues like Rebecca Nelson, Kaveh Waddell, Nora Caplan-Bricker, Nora Kelly, Emma Roller, Dustin Volz, Dylan Scott, Emily Schultheis, Andrew McGill, Clare Foran, Tim Alberta, Shane Goldmacher, and Caroline Mimbs-Nyce, was incredible. I wish we could have done it forever and it ended differently.

This project, from the magazine article to the proposal to the actual book, spanned across two presidential elections, one midterm election, and four jobs. So special thanks to the leadership, staff, and my colleagues at all of them, particularly John Helton, Ed Timms, Kris Viesselman, Jason Dick, Griffin Connolly, Bridget Bowman, Alex Gangitano, Gillian Roberts, Alex Roarty, Clyde McGrady, Chris Hale, Pablo Manriquez, Camila DeChalus, and the late Steve Komarow at *Roll Call;* Bob Cusack and Ian Swanson, Kyle Balluck, Tristan Lejune, Jesse Byrnes, Rachel Frazin, Brooke Siepel, Alicia Cohn, Tal Axelrod, Chris Mills Rodrigo, and so many others at the *Hill,* and Mike Madden, Jacob Brogan, Mary Jo Murphy, Sophia Nguyen, Steve Levingston, Carlos Lozada, Chris Shea, David Swerdlick, and Adam Kushner at the *Washington Post.* Adam, I still heard your voice telling me to substantiate things in the back of my head when I was writing this book and it made me better.

I was first recruited to *National Journal* thanks to the mentorship of Ron Fournier, who wrote movingly about his son Tyler being on the spectrum. This book would not exist without him. Not only did he introduce me to Heather, he also dropped a book called *NeuroTribes,* written by a journalist at *Wired* magazine, on my desk and introduced me to Steve Komarow (God rest his soul), who recruited me to *Roll Call* after leaving *National Journal.* I owe him (and Tyler) more than I can ever say and hope to pay it forward.

That journalist from *Wired* who wrote *NeuroTribes,* Steve Silberman, was also a massive influence on this book from the moment I interviewed him for the initial magazine piece to my time reading his book (as fate would have it, while writing the piece, I met him on a train to New York and we became fast friends). I am so thankful Steve undertook this project and I am grateful for his boosting of my work and writing. His book provided the intellectual foundation for this one and I hope that it is a house worthy of the sturdy rock that he laid down.

The only people who know how painful writing a book can be sometimes are fellow authors, and so many were in my corner. So a special thanks to all of them who listened to me vent and fret about it and who gave useful encouragement or advice, not limited to Matt Lewis, Nicole Cliffe, Lyz Lenz, Jonathan Cohn, Tom LoBianco Patricia Affriol, Wesley Lowery, Jill Filipovic, Sarah Frier, Zach D. Carter, Dave Weigel, Adam Serwer (who suggested the Frederick Douglass speech that bookends the last chapter), and countless others. Their friendship and support were vital and necessary. In the same token, I want to thank those who took the time to read this book and give productive feedback. Simone Pathe was literally the first person to whom I said that I had signed a book deal since we were working together at *Roll Call* and I walked out of the phone room. Her critiques and edits of each chapter made them sharper and more pointed. Similarly, much thanks to Felicia Sonmez, Zach, and Daniel Wiser for all being willing to read the manuscript to let me know how this could be improved upon. I am one of the most fortunate writers and human beings I know.

This book would not exist without the support of my dearest and oldest friends. JohnPaul Trotter has been my oldest friend for twenty years and he spent many a night on the phone with me when I was nervous about how this book would turn out. He is my brother in more ways than one. Tim Jewell and Geoff Sabir are dear partners whom I've known since high school. As an autistic person, I value having some degree of certainty and sameness in my life and that has been amply provided to me by many of the same friends I have had since college and many other friends I have picked up since then, primarily Maddy Will (who hired me at the *Daily Tar Heel* all those years ago when we were students at UNC), Joe Lenahan, Mary Tyler March, Claire Williams, Daniel Schere, Andy Willard, Sarah Brown, Russell Paige, Tara, Aaron, and Susanna Payne, Nicole, and Paige Comparato. In addition, thanks to Kristin Herbruck for providing an excellent headshot and plenty of cat videos when I was in the depths of writing.

Last, this book is for all the autistic students who messaged in the years since I started writing about autism who want to get into journalism. I want you to know you too have a space in media and a story that deserves to be told. And you—yes, you—can write a book too. We need you, and I need you, and I know you will build upon the things I got right and correct where I fell short.

Notes

Introduction

page

ix *Donald Trump in 2016:* Aaron Bycoffe, "Tracking Congress in the Age of Trump," FiveThirtyEight.com, January 30, 2017, https://projects.fivethirtyeight.com/congress-trump-score/cheri-bustos/.

x *but not flatly denying:* Eric Garcia, "Democratic Young Guns Take Different Approach to Rebuilding Party," *Roll Call,* October 11, 2017, https://www.rollcall.com/2017/10/11/democratic-young-guns-take-different-approach-to-rebuilding-party/.

xi "Autistic Disturbances of Affective Contact": Leo Kanner, "Autistic Disturbances of Affective Contact," *Journal of Psychopathology, Psychotherapy, Mental Hygiene, and Guidance of the Child* (1943): 217–50, http://www.neurodiversity.com/library_kanner_1943.pdf.

"there are very few really warm-hearted fathers and mothers": Kanner, "Autistic Disturbances," 250.

"*kept neatly in a refrigerator*": "Medicine: Frosted Children," *Time,* April 26, 1948, http://content.time.com/time/subscriber/article/0,33009,798484,00.html.

who further smeared parents: Charles Rycoft, "Lost Children," *New York Review of Books,* May 4, 1967, https://www.nybooks.com/articles/1967/05/04/lost-children/.

Kanner would later "acquit" parents: Steve Silberman, *NeuroTribes: The Legacy of Autism and the Future of Neurodiversity* (New York: Random House, 2016), 301.

xii *the dustbin of pseudoscience:* John J. Pitney Jr., *The Politics of Autism* (Lanham, MD: Rowman and Littlefield, 2015), 20.

the idea that mercury in vaccines causes autism: Silberman, *NeuroTribes,* 334.

broadcast on CNN: "Jenny McCarthy's Autism Fight," CNN.com, April 8, 2008, archives.cnn.com/TRANSCRIPTS/0804/02/lkl.01.html.

Trump, being Trump, doubled down: Ryan Teague Beckwith, "Read the Full Text of the

Second Republican Debate," *Time,* September 16 2015, time.com/4037239/second-republican-debate-transcript-cnn/.

gave Trump a pass: Beckwith, "Read the Full Text."

xiii *a major essay:* Eric Garcia, "I'm Not Broken," *Atlantic,* December 4, 2015, www.theatlantic.com/politics/archive/2015/12/im-not-broken/446550/.

who killed twenty-six people: Ella Torres, "Sandy Hook Shooting: Victims of Gun Violence Commemorate 7th Anniversary of Massacre," *ABC News,* December 14, 2019, abcnews.go.com/US/sandy-hook-shooting-victims-gun-violence-commemorate-7th/story?id=67735556.

jumped 130 percent: Bonnie Rochman, "Guilt by Association: Troubling Legacy of Sandy Hook May Be Backlash Against Children with Autism," *Time,* December 19, 2012, healthland.time.com/2012/12/19/guilt-by-associationtroubling-legacy-of-sandy-hook-may-be-backlash-against-children-with-autism/.

"*on the autism scale":* Dylan Byers, "Scarborough: Holmes 'on Autism Scale,'" *Politico,* July 23, 2012, www.politico.com/blogs/media/2012/07/scarborough-holmes-on-autism-scale-129779.

Scarborough has a son: Drew Katchen, "Father Learns to Understand, Embrace His Son's Asperger's Syndrome," MSNBC.com, December 4, 2012, www.msnbc.com/morning-joe/father-learns-understand-embrace-his-son-msna16548.

He later apologized: Tommy Christopher, "Autistic Activist to Joe Scarborough: 'I Am Not a Murderer,'" Mediaite.com, July 24, 2012, https://www.mediaite.com/tv/joe-scarborough-on-autism-remarks-perhaps-i-could-have-made-my-point-more-eloquently/.

xiv "*Families Against Autistic Shooters":* Andrew Solomon, "Opinion: The Myth of the 'Autistic Shooter,'" *New York Times,* October 12, 2015, www.nytimes.com/2015/10/12/opinion/the-myth-of-the-autistic-shooter.html.

there is no evidence: David S. Im, "Template to Perpetrate," *Harvard Review of Psychiatry* 24, no. 1 (2016): 14–35, www.ncbi.nlm.nih.gov/pmc/articles/PMC4710161/, 10.1097/hrp.0000000000000087.

autism and violent behavior: Santhana Gunasekaran and Eddie Chaplin, "Autism Spectrum Disorders and Offending," *Advances in Mental Health and Intellectual Disabilities* 6, no. 6 (November 2012): 308–13, https://www.emerald.com/insight/content/doi/10.1108/20441281211285955/full/html.

speculation and misunderstanding: Solomon, "The Myth of the 'Autistic Shooter.'"

autism alone is never the cause: Deanna Pan, "The Media's Post-Newtown Autism Fail," *MotherJones,* December22,2012,www.motherjones.com/politics/2012/12/journalism-newtown-autistics/.

75 percent of all research: Office of Autism Research Coordination, National Institute of Mental Health, Autistica, Canadian Institutes of Health Research, and Macquarie University, "2016 International Autism Spectrum Disorder Research Portfolio Analysis Report," October 2019, 42, https://iacc.hhs.gov/publications/international-portfolio-analysis/2016/portfolio_analysis_2016.pdf.

But autism likely can't be cured: Mayo Clinic Staff, "Autism Spectrum Disorder—Diagnosis and Treatment—Mayo Clinic," Mayo Clinic, January 6, 2018, www.mayoclinic.org/diseases-conditions/autism-spectrum-disorder/diagnosis-treatment/drc-20352934.

xv *An estimated one in fifty-four children:* "Data & Statistics on Autism Spectrum Disorder," Centers for Disease Control and Prevention, November 15, 2018, www.cdc.gov/ncbddd/autism/data.html.

CDC study estimated that 2.2 percent of the adult population: Maggie Fox, "First US Study of Autism in Adults Estimates 2.2% Have Autism Spectrum Disorder," CNN.com, May 11, 2020, www.cnn.com/2020/05/11/health/autism-adults-cdc-health/index.html.

"overcoming": Mike Suriani, "Penny Hardaway Praises Son Jayden as 'Walking Miracle' for Overcoming Autism," WREG.com, February 5, 2020, wreg.com/news/penny-hardaway-praises-son-jayden-as-walking-miracle-for-overcoming-autism/.

achieving great things: Amber Payne, "Autistic Brothers Walk Tall in Southern University Marching Band," *NBC News,* November 29, 2015, www.nbcnews.com/news/nbcblk/autistic-brothers-walk-tall-southern-university-marching-band-n470631.

xvi *Rebecca Cokley coined the term:* Rebecca Cokley, "Reflections from an ADA Generation, Rebecca Cokley, TEDx University of Rochester," YouTube, July 25, 2018, www.youtube.com/watch?v=nmDk6ZE3npY.

xvii autism spectrum disorders (ASD): Susan L. Hyman, "New *DSM-5* Includes Changes to Autism Criteria," *AAP News,* June 2013, https://www.aappublications.org/content/early/2013/06/04/aapnews.20130604-1.

previously existed under various terms: Jessica Wright, "DSM-5 Redefines Autism," *Spectrum,* May 21, 2013, https://www.spectrumnews.org/opinion/dsm-5-redefines-autism/.

xviii *"extreme male brain" theory:* Hannah Furfaro, "The Extreme Male Brain, Explained," *Spectrum News,* May 1, 2019, https://www.spectrumnews.org/news/extreme-male-brain-explained/.

1. "Don't Let Me Be Misunderstood"

2 *came out as autistic:* Ashley Goudeau, "State Rep. Briscoe Cain Opens Up About Being on the Autism Spectrum," Kvue.com, April 25, 2019, https://www.kvue.com/article/news/politics/texas-legislature/state-rep-briscoe-cain-opens-up-about-being-on-the-autism-spectrum/269-bc913b8a-9d20-48c5-9e59-db80b3937745.

Democrat Yuh-Line Niou won her seat: Leah Carroll, "If You Think Cuomo Is Going to Save New York, Meet Legislator Yuh-Line Niou Who Respectfully Disagrees," Refinery29, May 4, 2020, https://www.refinery29.com/en-us/2020/05/9679194/yuh-line-niou-chinatown-nyc-coronavirus.

that same year: Rory Mondshein, "We Put the 'Able' in 'Disabled': Local Politician, Yuh-Line Niou, on Autism Spectrum Disorder," *Political Student,* June 6, 2016, http://thepoliticalstudent.com/2016/06/we-put-the-able-in-disabled-local-politician-yuh-line-niou-on-autism-spectrum-disorder/.

she tells the crowd: Reporting by the author, February 9, 2020.

3 *tells me later:* Jessica Benham, interview by the author, February 9, 2020.

a symptom of childhood schizophrenia: Bonnie Evans, "How Autism Became Autism," *History of the Human Sciences* 26 (3): 3–31, https://doi.org/10.1177/0952695113484320.

"*schizophrenic reactions occurring before puberty":* Ellen Herman, "DSM-I (1952)," *Autism History Project,* 2019, https://blogs.uoregon.edu/autismhistoryproject/topics/autism-in-the-dsm/dsm-i-1952/.

symptom of prepuberty schizophrenia: Ellen Herman, "DSM-II (1968)," *Autism History*

Project, 2019, https://blogs.uoregon.edu/autismhistoryproject/topics/autism-in-the-dsm/.

"*onset before 30 months of age*": Ellen Herman, "'DSM-III (1980),'" *Autism History Project,* 2019, https://blogs.uoregon.edu/autismhistoryproject/topics/autism-in-the-dsm/dsm-iii-1980/.

"*childhood onset pervasive developmental disorder*": Bryan H. King et al., "Update on Diagnostic Classification in Autism," *Current Opinion in Psychiatry* 27 (2): 105–9, https://doi.org/10.1097/yco.0000000000000040.

4 *"pervasive developmental disorder not otherwise specified":* Lina Zeldovich, "The Evolution of 'Autism' as a Diagnosis, Explained," *Spectrum News,* May 29, 2018, https://www.spectrumnews.org/news/evolution-autism-diagnosis-explained/.

but the criteria are not met: James Coplan, *Making Sense of Autistic Spectrum Disorders: Create the Brightest Future for Your Child with the Best Treatment Options* (New York: Bantam Books, 2010), 374.

"Autistic Disorder": Ellen Herman, "Diagnostic Criteria for Autistic Disorder," *Autism History Project,* 2019, https://blogs.uoregon.edu/autismhistoryproject/topics/autism-in-the-dsm/dsm-iii-r-1987/.

number of symptoms from eight to six: Ellen Herman, "'DSM-IV' (1994) and 'DSM-IVR' (2000)," *Autism History Project,* 2019, https://blogs.uoregon.edu/autismhistoryproject/topics/autism-in-the-dsm/dsm-iv-1994-and-dsm-ivr-2000/.

Hans Asperger: Silberman, *NeuroTribes,* 82–139.

existed on a "continuum": Elon Green, "Rewriting Autism History," *Atlantic,* August 17, 2015, https://www.theatlantic.com/health/archive/2015/08/autism-history-aspergers-kanner-psychiatry/398903/.

argued that autism existed on a spectrum: Christopher Gilberg, "Lorna Wing OBE, MD, FRCPsych Formerly Psychiatrist and Physician, Social Psychiatry Unit, Institute of Psychiatry, King's College London, Co-Founder of the UK National Autistic Society," *BJPsych Bulletin* 39 (1): 52–53, https://doi.org/10.1192/pb.bp.114.048900.

even referred children to a clinic: Herwig Czech, "Hans Asperger, National Socialism, and 'Race Hygiene' in Nazi-Era Vienna," *Molecular Autism* 9, no. 1 (April 19, 2018), https://doi.org/10.1186/s13229-018-0208-6.

translated Asperger's work: Silberman, *NeuroTribes,* 349.

her seminal 1981 article: Lorna Wing, "Asperger's Syndrome: A Clinical Account," *Psychological Medicine* 11, no. 1 (February 1981): 115–29, https://doi.org/10.1017/s0033291700053332.

restrictive and repetitive behaviors: Herman, "'DSM-IV' (1994) and 'DSM-IVR' (2000)."

5 *autism spectrum disorder:* Hyman, "New *DSM-5* Includes Changes to Autism Criteria."

"*restricted, repetitive patterns of behavior*": Herman, "Autism in the DSM, 1952–2013."

a surge in autism diagnoses: Morton Gernsbacher, Michelle Dawson, and H. Hill Goldsmith, "Three Reasons Not to Believe in an Autism Epidemic," *Current Directions in Psychological Science* (April 1, 2005), https://pubmed.ncbi.nlm.nih.gov/25404790/.

people who went undetected in the past: Judith S. Miller et al., "Autism Spectrum Disorder Reclassified: A Second Look at the 1980s Utah/UCLA Autism Epidemiologic Study," *Journal of Autism and Developmental Disorders* 43, no. 1 (June 13, 2012): 200–210, https://doi.org/10.1007/s10803-012-1566-0.

Americans with Disabilities Act: George H. W. Bush, "Remarks by President George

H. W. Bush at the ADA Signing Ceremony," U.S. Department of Justice, Civil Rights Division, July 26, 1990, https://www.ada.gov/ghw_bush_ada_remarks.html.

barely mentioned autism during deliberations: Pitney, *The Politics of Autism,* 24.

6 *the Individuals with Disabilities Education Act:* Pitney, *The Politics of Autism,* 24.

"establish autism definitively as a developmental disability": Pitney, *The Politics of Autism,* 24.

IDEA mandated that students with disabilities: "Sec. 300.101 Free Appropriate Public Education (FAPE)," Individuals with Disabilities Education Act, U.S. Department of Education, May 3, 2017, https://sites.ed.gov/idea/regs/b/b/300.101.

schools had to report the number of autistic students they served: Gernsbacher, Dawson, Goldsmith, "Three Reasons Not to Believe in an Autism Epidemic," 4–5.

7 *crawled up the steps:* Lauren Lantry, "On 30th Anniversary of Disability Civil Rights Protest, Advocates Push for More," *ABC News,* March 12, 2015, https://abcnews.go.com/US/30th-anniversary-disability-civil-rights-protest-advocates-push/story?id=69491417.

8 *"as many as one in one hundred and sixty-six children are diagnosed":* "Steven Tyler Joe Perry Autism Ad," 2008, YouTube/AeroCarol, https://www.youtube.com/watch?v=-fMIVRYKXfg.

9 *the regalia of studs and leather:* Alexis Petridis, "Judas Priest's Rob Halford: 'I've Become the Stately Homo of Heavy Metal,'" *Guardian,* July 3, 2014, https://www.theguardian.com/music/2014/jul/03/judas-priest-rob-halford-quentin-crisp-interview-redeemer-of-souls.

10 *cofounded the Congressional Autism Caucus:* Pitney, *The Politics of Autism,* 131.

"they needed help": Mike Doyle, interview with the author, February 9, 2020.

founded the organization Autism Speaks: Jane Gross and Stephanie Strom, "Autism Debate Strains a Family and Its Charity," *New York Times,* June 18, 2007, https://www.nytimes.com/2007/06/18/us/18autism.html.

"dedicated to funding global biomedical research": Debra Muzikar, "Autism Speaks Revamps Its Mission Statement," *Art of Autism,* October 15, 2016, https://the-art-of-autism.com/autism-speaks-revamps-its-mission-statement/.

quickly gained national prominence: Pitney, *The Politics of Autism,* 31.

"*I am proud to sign this bill into law":* George W. Bush "President's Statement on Combating Autism Act of 2006," White House Archives, December 19, 2006, https://georgewbush-whitehouse.archives.gov/news/releases/2006/12/20061219-1.html.

11 *many of his subjects lacked "warm-hearted" parents:* Kanner, "Autistic Disturbances," 250.

"*lofty abstractions that little room is left":* Leo Kanner, "Problems of Nosology and Psychodynamics of Early Infantile Autism," *American Journal of Orthopsychiatry* 19, no. 3 (1949): 422, https://doi.org/10.1111/j.1939-0025.1949.tb05441.x.

he'd observed only one mother fully embrace her son: Kanner, "Problems of Nosology," 422.

"*They were kept neatly in refrigerators which did not defrost":* Kanner, "Problems of Nosology," 425.

further popularized the concept: Silberman, *NeuroTribes,* 207.

"*I state my belief":* Roy Richard Grinker, *Unstrange Minds: Remapping the World of Autism* (New York: Basic, 2007).

"a state of mind": Bruno Bettelheim, *The Empty Fortress: Infantile Autism and the Birth of the Self* (New York: Free Press, 1972), 68.
"helped keep autism off the agenda": Pitney, *The Politics of Autism,* 20.
thus didn't require broader policy solutions: Pitney, *The Politics of Autism,* 20.
absolve parents of blame: Silberman, *NeuroTribes,* 301.

12 *the Developmental Disabilities Act:* Ruth Christ Sullivan, "The Politics of Definitions: How Autism Got Included in the Developmental Disabilities Act," *Journal of Autism and Developmental Disorders* 9, no. 2 (1979): 221–31, https://doi.org/10.1007/bf01531536.
"persons with developmental disabilities": "Developmental Disabilities Assistance and Bill of Rights Act," October 4, 1975, https://uscode.house.gov/statutes/pl/94/103.pdf.
"the establishment of many services for our children": Sullivan, "The Politics of Definitions," 221.
Defeat Autism Now!: Silberman, *NeuroTribes,* 334.
the use of vitamin B_6: Benedict Carey, "Bernard Rimland, 78, Scientist Who Revised View of Autism, Dies," *New York Times,* November 28, 2006, http://nytimes.com/2006/11/28/obituaries/28rimland.html.
intravenous use of the hormone secretin: Rosie Mestel, "Parents Look to Unproven Therapies to Solve Autism," *Los Angeles Times,* March 12, 2001, https://www.latimes.com/archives/la-xpm-2001-mar-12-he-36611-story.html.
"autism world has trailed": Ari Ne'eman, interview with the author, 2019.

13 *"as long as he lives":* Joanne Kaufman, "Ransom-Note Ads About Children's Health Are Canceled," *New York Times,* December 20, 2007, https://www.nytimes.com/2007/12/20/business/media/20child.html.
ditched the toxic idea: Joseph P. Shapiro, *No Pity: People with Disabilities Forging a New Civil Rights Movement* (New York: Three Rivers, 1994), 21–225.
"You had a greater willingness to support institutionalization": Ari Ne'eman, interview with the author, 2019.
examined coverage of autism: Clarke, Juanne Nancarrow, "Representations of Autism in US Magazines for Women in Comparison to the General Audience," *Journal of Children and Media,* Volume 6, Number 2, 2011 182–97, https://doi.org/10.1080/17482798.2011.587143.

14 *drugged and stabbed repeatedly:* "Film Provides Glimpse into Life of Autistic Teen Killed by His Mother," *CBS News,* August 30, 2013, https://www.cbsnews.com/news/film-provides-glimpse-into-life-of-autistic-teen-killed-by-his-mother/.
pleaded guilty to involuntary manslaughter: John Garcia, "Mother, Godmother Who Killed Autistic Teen Released from Prison," ABC7 Chicago, December 14, 2016, https://abc7chicago.com/news/mother-godmother-who-killed-autistic-teen-released-from-prison/1656353/.

15 *graduated electronic decelerator (GED):* Patricia M. Rivera et al., "Use of Skin-Shock at the Judge Rotenberg Center," http://www.effectivetreatment.org/using_graduated.html.
wrote that the shocks violated the UN's conventions: Juan Méndez, "Report of the Special Rapporteur on Torture and Other Cruel, Inhuman or Degrading Treatment or Punishment," United Nations Human Rights Office of the High Commissioner, March 4, 2013, https://www.ohchr.org/Documents/HRBodies/HRCouncil/RegularSession/Session22/A.HRC.22.53.Add.4_Advance_version.pdf.
Matthew Israel, the center's founder: Jen Quraishi, "'School of Shock' Founder Forced to

Resign," *Mother Jones,* May 27, 2011, https://www.motherjones.com/politics/2011/05/judge-rotenberg-forced-resign-school-shocks/.

agreed to five years of probation: Russell Contreras, "Charges Settled in Prank-Call Shock Therapy Case," Boston.Com, May 25, 2011, http://archive.boston.com/news/local/massachusetts/articles/2011/05/25/charges_settled_in_prank_call_shock_therapy_case/.

"present an unreasonable and substantial risk": "FDA Takes Rare Step to Ban Electrical Stimulation Devices for Self-Injurious or Aggressive Behavior," U.S. Food and Drug Administration, March 4, 2020, https://www.fda.gov/news-events/press-announcements/fda-takes-rare-step-ban-electrical-stimulation-devices-self-injurious-or-aggressive-behavior.

a parent named Eddie Sanchez: Paul Kix, "The Shocking Truth," *Boston,* June 17, 2008, https://www.bostonmagazine.com/2008/06/17/the-shocking-truth/.

16 *Bernard Rimland praised:* Silberman, *NeuroTribes,* 419.

measles-mumps-rubella vaccine: Pitney, *The Politics of Autism,* 25.

paid by attorneys representing parents of children: Pitney, *The Politics of Autism,* 51.

lost his medical license in 2010: Alice Park, "Doctor Behind Vaccine-Autism Link Loses License," *Time,* May 24, 2010, https://healthland.time.com/2010/05/24/doctor-behind-vaccine-autism-link-loses-license/.

Wakefield found an audience: Philip J. Hilts, "House Panel Asks for Study of a Vaccine," *New York Times,* April 7, 2000, https://www.nytimes.com/2000/04/07/us/house-panel-asks-for-study-of-a-vaccine.html.

shooting a cantaloupe: Mary Ann Akers, "Dan Burton, Protecting the House from Terrorists (Alone)," *Washington Post,* June 19, 2009, http://voices.washingtonpost.com/sleuth/2009/06/_rep_dan_burton_r-ind.html.

17 *Burton spoke anecdotally:* "Autism: Present Challenges, Future Needs—Why the Increased Rates?," U.S. Government Printing Office, 2001, https://www.govinfo.gov/content/pkg/CHRG-106hhrg69622/html/CHRG-106hhrg69622.htm.

warned that this could cause a dangerous drop: Hilts, "House Panel Asks for Study of a Vaccine."

Waxman told me in 2019: Henry Waxman, interview with the author, 2019.

hearing on autism: "Funding Autism Research," C-SPAN, August 5, 2009, https://www.c-span.org/video/?288235-1/funding-autism-research.

he delivered a floor speech in sign language: Tom Harkin, "Senator Tom Harkin American Sign Language Senate Speech," Google Arts and Culture, July 13, 1990, https://artsandculture.google.com/asset/senator-tom-harkin-american-sign-language-senate-speech-senator-tom-harkin/zwFD2G4XUFGIPA?hl=en.

18 "*I went through a period*": Tom Harkin, interview with the author, 2018.

"*the Mercurys*": Jane Gross and Stephanie Strom, "Autism Debate Strains a Family and Its Charity," *New York Times,* June 18, 2007, https://www.nytimes.com/2007/06/18/us/18autism.html.

quit the organization in 2009: Meredith Waldman, "Autism's Fight for Facts: A Voice for Science," *Nature* 479 (2011): 28–31, https://doi.org/10.1038/479028a.

"*We've seen just a skyrocketing autism rate*": Glenn Kessler, "Fact Checking Statements on Vaccines, Including Obama in 2008," *Washington Post,* February 5, 2015, https://www.washingtonpost.com/news/fact-checker/wp/2015/02/05/fact-checking-statements-on-vaccines-including-obama-in-2008/.

19 "*We're still unable to cure diseases like Alzheimer's or autism*": Barack Obama, "Re-

marks by the President on the BRAIN Initiative and American Innovation," Obama White House, April 2, 2013, https://obamawhitehouse.archives.gov/the-press-office/2013/04/02/remarks-president-brain-initiative-and-american-innovation.

delivered an address titled: Jim Sinclair, "Don't Mourn for Us," presented at the International Conference on Autism, 1993, https://www.autreat.com/dont_mourn.html.

20 *Autreat, a getaway for autistic people:* Silberman, *NeuroTribes,* 448–50.

was one of the first memoirs by an autistic adult: Sarah Pripas-Kapit, "Historicizing Jim Sinclair's 'Don't Mourn for Us': A Cultural and Intellectual History of Neurodiversity's First Manifesto," *Autistic Community and the Neurodiversity Movement* (November 2019), 23–39, https://doi.org/10.1007/978-981-13-8437-0_2.

laid the groundwork for future advocates: Jim Sinclair, "Being Autistic Together," *Disability Studies Quarterly* 30, no. 1 (2010): https://doi.org/10.18061/dsq.v30i1.1075.

21 *he was reaching out to groups:* Ari Ne'eman, interview with the author, 2019.

had petitioned the United Nations: Robert McRuer and Anna Mollow, *Sex and Disability* (Durham, NC: Duke University Press, 2012), 279.

"The way I saw the work": Ari Ne'eman, interview with the author, 2019.

as its slogan: "Nothing About Us Without Us! Who We Are & What We Do," Autistic Self Advocacy Network, April 2019, https://autisticadvocacy.org/wp-content/uploads/2019/04/ASAN-flyer-1-pager-2018.pdf.

firmly declaring its goals: James I. Charlton, *Nothing About Us Without Us* (Berkeley: University of California Press, 1998).

successfully coordinated with prominent disability rights activists: Silberman, *NeuroTribes,* 462.

22 *own college campuses:* "Autism Campus Inclusion," Autistic Self Advocacy Network, 2020, https://autisticadvocacy.org/projects/education/aci/.

was appointed by President Barack Obama: "Ari Ne'eman," Healthpolicy.Fas.Harvard.Edu, 2020, https://healthpolicy.fas.harvard.edu/people/ari-neeman.

and was a public member: "Leadership," Autistic Self Advocacy Network, 2020, https://autisticadvocacy.org/about-asan/leadership/.

Combating Autism Act: Interagency Autism Coordinating Committee, 2019, "IACC Legislation," https://iacc.hhs.gov/about-iacc/legislation/.

aggressive campaign: "#StopCombatingMe," Autistic Self Advocacy Network, 2014, https://autisticadvocacy.org/resources/stopcombatingme/.

was reauthorized as the Autism CARES: John J. Pitney, Jr., "Autism and Accountability," presented at the Prepared for delivery at the 2020 Annual Meeting of the American Political Science Association, September 11, 2020, https://www.scribd.com/document/475658529/Autism-and-Accountability#download.

guest-edited an issue of Wired *magazine:* "President Obama Guest Edits *Wired,*" *Wired,* October 2016, https://www.wired.com/magazine/president-obama-guest-edits-wired/.

"They might be on the spectrum": Scott Dadich, "Barack Obama Talks AI, Robo Cars, and the Future of the World," *Wired,* October 12, 2016, https://www.wired.com/2016/10/president-obama-mit-joi-ito-interview/.

23 *tweeted his support of Trump:* Bob Wright, "Go Donald Trump. The Best Choice for All Republicans and All Americans!," Twitter, April 18, 2016, https://twitter.com/bobwrightnbc/status/722223562900299778.

tried to persuade Wright: Bob Wright and Dianne Mermigas, *The Wright Stuff: From NBC to Autism Speaks* (New York: Rosetta Books, 2016).

scared away NBC: Eliot Brown, "Remember Trump City?," *Observer,* August 5, 2008, https://observer.com/2008/08/remember-trump-city/.

first aired, in 2004: Wright and Mermigas, *The Wright Stuff,* 145–46.

"bitterly disappointed": Wright and Mermigas, *the Wright Stuff,* 313.

Trump did so his first year: Donald Trump, "President Donald J. Trump Proclaims April 2, 2017, as World Autism Awareness Day," White House, March 31, 2017, https://www.whitehouse.gov/presidential-actions/president-donald-j-trump-proclaims-april-2-2017-world-autism-awareness-day/.

angered many autistic self-advocates: Oliver-Ash Kleine (since came out as non-binary and changed name), "The White House Turned Blue for 'Autism Awareness,' That's Actually Bad for Autistic People," *Mother Jones,* April 17, 2017, https://www.motherjones.com/politics/2017/04/trumps-autism-awareness-effort-not-show-support-autistic-people/.

none of its board members were openly autistic: Melissa Dahl, "A Leading Autism Organization Is No Longer Searching for a 'Cure,'" *Cut,* October 18, 2016, https://www.thecut.com/2016/10/autism-speaks-is-no-longer-searching-for-a-cure.html.

"I Am Autism": Claudia Wallis, "'I Am Autism': An Advocacy Video Sparks Protest," *Time,* November 6, 2009, http://content.time.com/time/health/article/0,8599,1935959,00.html.

"I know where you live": "I Am Autism Commercial by Autism Speaks," YouTube video, YouTube/Find Yaser, https://www.youtube.com/watch?v=9UgLnWJFGHQ.

24 *a piece on the nonprofit's website:* Suzanne Wright, "Autism Speaks to Washington—A Call for Action," Internet Archive, November 12, 2013, https://web.archive.org/web/20131112033720/https://www.autismspeaks.org/news/news-item/autism-speaks-washington-call-action.

John Elder Robison: John Elder Robison, interview with the author, 2015.

Trump reportedly met with Andrew Wakefield: Zack Kopplin, "Trump Met with Prominent Anti-Vaccine Activists During Campaign," *Science,* November 18, 2016, https://www.sciencemag.org/news/2016/11/trump-met-prominent-anti-vaccine-activists-during-campaign.

"*there will be no single 'cure'":* "For the Record," *Autism Speaks,* 2020, https://www.autismspeaks.org/record.

the first two autistic members: Dahl, "A Leading Autism Organization."

vice president of the organization: Valerie Paradiz, "One Woman's Journey Navigating Autism and Mental Health," *Autism Speaks,* July 8, 2019, https://www.autismspeaks.org/blog/one-womans-journey-navigating-autism-and-mental-health.

25 *criticized Ne'eman:* Amy Harmon, "Nominee to Disability Council Is Lightning Rod for Dispute on Views of Autism," *New York Times,* March 27, 2010, https://nytimes.com/2010/03/28/health/policy/28autism.html.

"want to be associated with the autism spectrum": Wright and Mermigas, *The Wright Stuff,* 366.

right of the middle on the opposite end: Lydia X. Z. Brown, interview with the author, 2018.

26 *Dr. Sanjay Gupta profiled Mel Baggs:* Sanjay Gupta, "Behind the Veil of Autism,"

CNN, February 20, 2007, http://edition.cnn.com/HEALTH/blogs/paging.dr.gupta/2007/02/behind-veil-of-autism.html.

video on YouTube called In my Language: Mel Baggs, "In My Language," YouTube, 2007, https://www.youtube.com/watch?v=JnylM1hI2jc.

University of California, Berkeley: Hari Srinivasan, interview with the author, 2020.

Jessica Benham got into autism advocacy: Jessica Benham, interview with the author, 2019.

27 *"It's certainly a milestone":* Jessica Benham, interview with the author, 2020.

28 *Benham has won their endorsement:* Hannah Lynn, "Pittsburgh LGBTQ Political Advocacy Group Announces Primary Election Endorsements," *Pittsburgh City Paper,* February 10, 2020, https://www.pghcitypaper.com/pittsburgh/pittsburgh-lgbtq-political-advocacy-group-announces-primary-election-endorsements/Content?oid=16731149.

she won her primary: Julia Terruso, "Bernie Sanders Is Done but His Fans in Pa. Keep Winning Primaries," *Philadelphia Inquirer,* June 16, 2020, https://www.inquirer.com/news/nikil-saval-rick-krajewski-progressive-2020-pennsylvania-primary-harrisburg-20200616.html.

2. "In My Mind, I'm Going to Carolina"

29 *about the decline of the coal industry:* Kris Maher, "In Pro-Trump West Virginia Coal Country, the Jobs Keep Leaving," *Wall Street Journal,* October 28, 2019, https://www.wsj.com/articles/in-pro-trump-west-virginia-coal-country-the-jobs-keep-leaving-11572269967.

the gripping poverty: Aimee Picchi, "West Virginia Poverty Gets Worse Under Trump Economy, Not Better," *CBS News,* September 28, 2018, https://www.cbsnews.com/news/west-virginia-poverty-gets-worse-under-trump-economy-not-better/.

41.5 deaths per 100,000 people: "Harm Reduction Program," Office of Epidemiology and Prevention Services, 2020, https://oeps.wv.gov/harm_reduction/pages/default.aspx.

why the state had shifted: Mike Kelly, "A Town in Opioids' Grip Still Looks to President Trump with Hope," North Jersey Media Group, October 19, 2018, https://www.northjersey.com/story/news/columnists/mike-kelly/2018/10/17/town-opioids-grip-looks-donald-trump-hope/1381268002/.

voting for Democrats in the twentieth century: Philip Bump, "How West Virginia Explains Donald Trump's Political Success," *Washington Post,* February 9, 2016, https://www.washingtonpost.com/news/the-fix/wp/2016/02/09/how-west-virginia-explains-donald-trumps-political-success/.

the University of Alabama: Megan Davis, interview with the author, 2018.

30 *Ruth Christ Sullivan:* Carter Taylor Seaton, "The Pioneer," *Huntington Quarterly,* September 27, 2018, https://huntingtonquarterly.com/2018/09/27/issue-79-the-pioneer/.

the cofounder of the National Society for Autistic Children: Pitney, *The Politics of Autism,* 20–21.

Sullivan was focused: Silberman, *NeuroTribes,* 304.

the movie Rain Man: Megan Osborne, "Autism Services Center Marks 40 Years with New Facility," *Herald-Dispatch,* May 27, 2019, https://www.herald-dispatch.com/news/autism-services-center-marks-40-years-with-new-facility/article_73fa9c28-1e98-5c01-b24e-97b316e374d4.html.

When Sullivan relocated to Huntington: Silberman, *NeuroTribes,* 299.

NSAC's Information and Referral Service: Kyla Asbury, "Autism Center Celebrates 30 Years," *Herald-Dispatch,* March 28, 2009, https://www.herald-dispatch.com/news/recent_news/autism-center-celebrates-years/article_ae278fb9-b9e6-5dc8-b905-7fa9fb55a7bd.html.

the first state to specifically include autism: Silberman, *NeuroTribes,* 299.

In 1984, Sullivan led efforts: Barbara Becker-Cottrill, "From the Director," The West Virginia Autism Training Center at Marshall University-News and Views, 2004, https://www.yumpu.com/en/document/read/15385274/2004-spring-issue-marshall-university.

it began offering services for autistic students: Rebecca Hansen, interview with the author, 2018.

Old Main Suite building: "Front Page," College Program for Students with Autism Spectrum Disorder, Marshall University, https://www.marshall.edu/collegeprogram/.

Richie Combs, who was in his final year of college: Richie Combs, interview with the author, 2018.

Zachariah Lewis: Zachariah Lewis, interview with the author, 2018.

32 *81 percent of college students with ASD:* Xin Wei et al., "Postsecondary Pathways and Persistence for STEM Versus Non-STEM Majors: Among College Students with an Autism Spectrum Disorder," *Journal of Autism and Developmental Disorders* 44, no. 5 (2014): 1159–67, https://doi.org/10.1007/s10803-013-1978-5.

33 *because it implies that disabled people are getting preferential treatment:* Morton Ann Gernsbacher et al., "'Special Needs' Is an Ineffective Euphemism," *Cognitive Research: Principles and Implications* 29, no. 1 (December 19, 2016), https://www.ncbi.nlm.nih.gov/pmc/articles/PMC5256467/.

34 *Timotheus Gordon:* Timotheus Gordon, interview with the author, 2018.

"*There's this element of disbelief*": M. Remi Yergeau, interview with the author, 2018.

Hari Srinivasan: Hari Srinivasan, "The Neurodiversity of Autism at UC Berkeley," *Daily Californian,* April 29, 2019, https://www.dailycal.org/2019/04/28/it-really-is-a-spectrum/.

35 *"my counselor knew most of the instructors":* Hari Srinivasan, interview with the author, 2020.

37 *"disability is a personal problem":* M. Remi Yergeau, interview with the author, 2018.

Berkeley's disabled students program: Hari Srinivasan, interview with the author, 2020.

who goes by Aria on social media: Aria, e-mail interview with the author, 2019.

38 *perceive as low-functioning:* Lee Romney, "Experts Called Hari 'Low-Functioning' for Years. Then He Found His Voice," KALW, June 11, 2019, https://www.kalw.org/post/experts-called-hari-low-functioning-years-then-he-found-his-voice#stream/0.

was even subjected to applied behavior analysis: Hari Srinivasan, interview with the author, 2020.

admitted his past experiences using accommodations: Finn Gardiner, interview with the author, 2018.

39 *Tufts University in Massachusetts:* "Finn Gardiner—Lurie Institute for Disability Policy," Heller School, Brandeis, 2020, https://heller.brandeis.edu/lurie/about/gardiner.html.

Lydia Brown: Lydia X. Z. Brown, "*Autistic Hoya,* 2020, https://www.autistichoya.com/.

taught a class at Tufts University: Lydia X. Z. Brown, interview with the author, 2018.

41 *Hansen, the director of campus-based services:* Rebecca Hansen, interview with the author, 2018.
executive functioning: Bonnie Glickman, "Executive Function Challenges in Children with Asperger Profiles," Asperger/Autism Network, July 21, 2016, https://www.aane.org/executive-function-challenges-children-asperger-profiles/.
42 *This is the case with many of the students at Marshall:* Rebecca Hansen, interview with the author, 2018.
Marshall teaches "adaptive living skills": "The WV Autism Training Center's College Program for Students with Autism Spectrum Disorder at Marshall University," Marshall University, https://www.marshall.edu/collegeprogram/files/College-Program-Summary-and-Overview.pdf.
Richie Combs told me: Richie Combs, interview with the author, 2018.
43 *Developing a Social Radar:* "The WV Autism Training Center's College Program for Students with Autism Spectrum Disorder at Marshall University," Marshall University, https://www.marshall.edu/collegeprogram/files/College-Program-Summary-and-Overview.pdf.
said many students receive assistance: Rebecca Hansen, interview with the author, 2018.
Gardner said it was important: Sara Gardner, interview with the author, 2018.
44 *overlap with Marshall's:* "Neurodiversity Navigators," Bellevue College, 2020, https://36d5l8225ig13rrnnc3w4af9-wpengine.netdna-ssl.com/wp-content/uploads/sites/68/2020/09/NdN-one-page-2021.pdf.
45 *most autistic college students, the major problem wasn't:* Scott L. J. Jackson et al., "Brief Report: Self-Reported Academic, Social, and Mental Health Experiences of Post-Secondary Students with Autism Spectrum Disorder," *Journal of Autism and Developmental Disorders* 48, no. 3 (2017): 643–50, https://doi.org/10.1007/s10803-017-3315-x.
47 *Savicki argued it was "much more competitive":* Andrew Savicki, "The Hot Seat: Oklahoma's 5th District," *Medium*, September 3, 2018, https://medium.com/@rudnicknoah/the-hot-seat-oklahomas-5th-district-afdfd74f087.
Horn proved Savicki right: Michael Burke, "Dem Kendra Horn Wins Oklahoma Seat in Major Upset," *Hill*, November 7, 2018, https://thehill.com/homenews/house/415413-dem-kendra-horn-wins-oklahoma-seat-in-major-upset.
"I'm just stuck in a never-ending loop": Andrew Savicki, interview with the author, 2019.
and himself a high school dropout: "John Elder Robison," *Psychology Today*, 2020, https://www.psychologytoday.com/us/experts/john-elder-robison.
48 "*What we need to balance that are successful adults":* John Elder Robison, "The Challenges of Neurodiversity in Colleges," *Look Me in the Eye*, October 13, 2015, http://jerobison.blogspot.com/2015/10/the-challenges-of-neurodiversity-in.html.
"*most of us can grow up to live independently":* Robison, "The Challenges of Neurodiversity."
Caudel, who did not know he was autistic: Dave Caudel, interview with the author, 2018.
49 *Bridge Program:* Dave Caudel, interview with the author, 2018.
which helps underrepresented students: "About the Program—Fisk-Vanderbilt Master's-to-PhD Bridge Program," Fisk-Vanderbilt Master's-to-PhD Bridge Program, https://www.fisk-vanderbilt-bridge.org/program.
who founded the bridge program: Spencer Turney, "Ph.D. Bridge Programs Show Success

in Fostering Diversity and Inclusion, According to New Study," Vanderbilt University, December 6, 2019, https://news.vanderbilt.edu/2019/12/06/ph-d-bridge-programs-show-success-in-fostering-diversity-and-inclusion-according-to-new-study/.

promote neurodiverse talent: "The Frist Center for Autism and Innovation," Vanderbilt University, 2020, https://my.vanderbilt.edu/autismandinnovation/.

"If we only take care of those wealthy kids who have autism": Dave Caudel, interview with the author, 2018.

50 *Vanderbilt Autism and Neurodiversity Alliance (VANA):* "Children's Specialized Hospital—CDID Advisory Board," Center for Discovery, Innovation and Development, 2020, http://www.cdid.org/advisory-board.

"The thing that keeps me up late at night": Dave Caudel, interview with the author, 2020.

"*There are two sides to Vanderbilt's treatment of neurodiversity":* Claire Barnett, "The Neurodiversity Movement," April 2, 2018, https://vanderbilthustler.exposure.co/the-neurodiversity-movement?embed=true.

Vice President Mike Pence's office: Claire Barnett, permission from the interviewee, 2020.

51 "*she would tell people, 'I got hired because I'm autistic'":* Dave Caudel, interview with the author, 2020.

3. "That Ain't Workin'"

53 *he didn't even intend to admit he was autistic:* John Marble, interview with the author, 2018.

54 *to work on Hillary Clinton's campaign:* "John Marble," Pivot Neurodiversity, 2020, https://www.pivotdiversity.com/john-marble.

Autism Advantage: "About Us," Neurodiversity Pathways, 2019, https://ndpathways.org/about-us/.

55 *individual service plans:* "Workplace Readiness Program," Neurodiversity Pathways, 2018, https://ndpathways.org/wp-content/uploads/2018/11/WPR-Brochure-Sept-2018-2.pdf.

Marble frequently took students around to companies: John Marble, interview with the author, 2018.

"*soft bigotry of low expectations":* Linda Wertheimer, "A Final Word with President's Faithful Speechwriter," NPR, June 21, 2006, https://www.npr.org/templates/story/story.php?storyId=5499701.

"*Silicon Valley of tomorrow":* Nancy Snyderman, "Standards for Diagnosing Autism under Review," *Today,* June 14, 2012, https://www.today.com/video/standards-for-diagnosing-autism-under-review-44548675739.

59 *when she pioneers new methods to handle livestock:* Temple Grandin, *Thinking in Pictures,* 2006, https://www.grandin.com/inc/visual.thinking.html.

learning to set off bombs: John Elder Robison, *Look Me in the Eye* (New York: Crown, 2007).

and then working on toys: Robison, *Look Me in the Eye,* 192–96.

before moving to cars: Robison, *Look Me in the Eye,* 232–36.

60 *commonly cited statistic:* Paul Austin and Andrew Williams, "As People on the Autism Spectrum Move into Adulthood, Who Will Hire Them?," *Smarter Business Review,*

March 26, 2019, https://www.ibm.com/blogs/services/2019/03/26/as-people-on-the-autism-spectrum-move-into-adulthood-who-will-hire-them/.

85 percent of college-educated autistic people: Nicole Lyn Pesce, "Most College Grads with Autism Can't Find Jobs. This Group Is Fixing That," *MarketWatch,* March 2, 2019, https://www.marketwatch.com/story/most-college-grads-with-autism-cant-find-jobs-this-group-is-fixing-that-2017-04-10-5881421.

"*the least common outcome*": Anne Roux el al., "2017 National Autism Indicators Report: Developmental Disability Services and Outcomes in Adulthood," A. J. Drexel Autism Institute, Drexel University, May 24, 2017, https://drexel.edu/autismoutcomes/publications-and-reports/publications/National-Autism-Indicators-Report-Developmental-Disability-Services-and-Outcomes-in-Adulthood/.

a quarter of autistic adults: Anne Roux el al., "2017 National Autism Indicators Report," 13.

hiring autistic people is not an act of charity: John Marble, interview with the author, 2018.

61 *regretted disclosing she was autistic to her employers:* Marcelle Ciampi, interview with the author, 2019.

the pen name Samantha Craft: Marcelle Ciampi, "Samantha," My Spectrum Suite, 2020, http://www.myspectrumsuite.com/keynote-speaker-workshop-presenter/.

for her book Everyday Aspergers: Samantha Craft, *Everyday Aspergers: A Journey on the Autism Spectrum* (Lancaster, UK: Explainer, 2018).

employers begin to question: Marcelle Ciampi, interview with the author, 2019.

"*Our goal is that this initiative encourages*": Dane Holmes, "Embracing Neurodiversity," Goldman Sachs, April 2, 2019, https://www.goldmansachs.com/careers/blog/posts/dane-holmes-embracing-neurodiversity.html.

62 "*traditional interview techniques*": H. Annabi et al., "Autism @ Work Playbook: Finding Talent and Creating Meaningful Employment Opportunities for People with Autism," Information School, University of Washington, 2019, https://disabilityin.org/wp-content/uploads/2019/07/Autism_At_Work_Playbook_Final_02112019.pdf.

robots out of Legos based on a set of instructions: Jeff Chu, "Why SAP Wants to Train and Hire Nearly 700 Adults with Autism," Inc.com, June 12, 2015, https://www.inc.com/jeff-chu/sap-autism-india.html.

these exercises was to gauge how candidates read and follow instructions: Chu, "Why SAP Wants to Train."

has pledged to create one million jobs for autistic people: Thorkil Sonne, "First Person: Founder of Specialisterne," *ABC News,* April 1, 2010, https://abcnews.go.com/WN/person-founder-specialisterne/story?id=10260011.

plenty of misconceptions: Tara Cunningham, interview with the author, 2018.

63 "*being honest and loyal*": Pesce, "Most College Grads with Autism."

he tweeted that disabled employees were not loyal: Steve Lieberman, 2019, Twitter, August 20, 2019, https://twitter.com/stevemlieberman/status/1163967410325610497.

"*I worry about saying, 'Hire an autistic person'*": Tara Cunningham, interview with the author, 2018.

64 *focusing on a certain type:* Julia Bascom, interview with the author, 2019.

65 *withdrew from the discussion:* Julia Bascom, correspondence with the author, 2020.

Kriss told me in 2020: Alan Kriss, interview with the author, 2020.

66 *"While it's good to advocate for decreased barriers":* endever star, interview with the author, 2018.

A total of 383,941 autistic people: Social Security Administration, Office of Retirement and Disability Policy, and Office of Research, Evaluation, and Statistics, 2020, "SSI Annual Statistical Report, 2019," Social Security, August 2020, https://www.ssa.gov/policy/docs/statcomps/ssi_asr/2019/ssi_asr19.pdf.

67 *Meanwhile, people who have never been able to work:* endever star, interview with the author, 2018.

"One of the times they told me I wasn't disabled enough": Andrew Savicki, interview with the author, 2019.

68 *President Franklin D. Roosevelt:* Howard D. Samuel, "Troubled Passage: The Labor Movement and the Fair Labor Standards Act," *Monthly Labor Review,* December 2000, https://www.bls.gov/opub/mlr/2000/12/art3full.pdf.

"whose earning or productive capacity is impaired": "From the New Deal to the Real Deal: Joining the Industries of the Future," 2018, Washington, DC: National Council on Disability, https://ncd.gov/sites/default/files/Documents/NCD_Deal_Report_508.pdf.

processed pieces of foam: Maxfield Sparrow, interview with the author, 2019.

championed by Republican legislators: Deborah Cohen, "Texas Took Great First Step in Ending Subminimum Wage," *Amarillo Globe-News,* June 28, 2019, https://www.amarillo.com/news/20190628/cohen-texas-took-great-first-step-in-ending-subminimum-wage.

69 *nonprofit organizations that employ disabled workers:* Christine Pulfrey, "Texas Offers Disabled Workers Minimum Wage on State Contracts," Bloomberg Tax, June 11, 2019, https://news.bloombergtax.com/payroll/texas-offers-disabled-workers-minimum-wage-on-state-contracts.

while he supported the authority: Senator Maggie Hassan, "Senator Hassan Presses Labor Nominee Acosta on Workplace Safety & Worker Protections," YouTube Video, March 22, 2017, https://www.youtube.com/watch?v=5G2S8EXlzNY.

"The idea that everyone with autism can achieve": "Vocational Options," National Council on Severe Autism, December 10, 2018, https://www.ncsautism.org/vocational-options.

"the disabled who have no voice": "NCSA Letter to Biden Campaign Re Disability Policy," National Council on Severe Autism, June 4, 2020, https://www.ncsautism.org/blog//ncsa-letter-to-biden-campaign-re-disability-policy.

"Voice of the Retarded": "Giving a Voice to Families and Guardians: A Survey of Families and Guardians of Individuals with Intellectual and Developmental Disabilities in Various Residential Settings," VOR, April 2015, https://www.vor.net/images/stories/pdf/VOR_Survey_Giving_a_Voice_to_Families_and_Guardians_April_2015.pdf.

"specialized wages": "The Movement to Eliminate Work Centers and 14(c) Wage Certificates," VOR, 2019, https://www.vor.net/images/stories/VOR_VOICE/Spring-2019.pdf.

often lack experience working with disability advocates: Julia Bascom, interview with the author, 2019.

"they're also the minority opinion": Julie Christensen, interview with the author, 2020.

70 *in states like Maine that have shuttered their programs:* NCSA, "Vocational Options."

who left sheltered workshops: Janet Phoenix and Tyler Bysshe, "Transitions: A Case Study of the Conversion from Sheltered Workshops to Integrated Employment in Maine," Milken Institute School of Public Health, George Washington University, July 2015, http://www.vaaccses.org/vendorimages/vaaccses/REPORT_Transitions_ConversionFromShelteredWorkshops_Maine_July2015.pdf.

saw 80 percent of people who worked in sheltered workshops: "Subminimum Wage and Supported Employment," 2012, National Council on Disability, https://ncd.gov/sites/default/files/NCD_Sub%20Wage_508.pdf.

Vermont has a 38 percent integrated-employment rate: Jean Winsor et al., "State Data: The National Report on Employment Services and Outcomes Through 2016," Institute for Community Inclusion, University of Massachusetts Boston, 2018, https://www.thinkwork.org/sites/default/files/files/narrative_accessible.pdf.

71 *double the national average of 19 percent:* Sarah Katz, "Biden's Disability Plan Could Close the Equal-Pay Loophole," *Atlantic,* August 10, 2020, https://www.theatlantic.com/politics/archive/2020/08/2020-election-subminimum-wage-disabilities/615085/.

"*wage-earning is not the primary purpose of these places*": NCSA, "Vocational Options."

Congress passed the Social Security Disability Amendments Act: Ari Ne'eman, "(Almost) Everything You Need to Know About Sheltered Workshops: Part 2 of 2," Sometimes a Lion, September 26, 2015, https://arineeman.com/2015/09/26/sheltered-workshops-part-2/.

"*It could be*": Julia Bascom, interview with the author, 2020.

72 *the Democratic and Republican Party platforms:* Michelle Diament, "Democrats, Republicans Urge End to Subminimum Wage," Disability Scoop, July 28, 2016, https://www.disabilityscoop.com/2016/07/28/democrats-republicans-subminimum/22548/.

Commission on Civil Rights: "Subminimum Wages: Impacts on the Civil Rights of People with Disabilities," U.S. Commission on Civil Rights, September 17, 2020, https://www.usccr.gov/pubs/briefing-reports/2020-09-17-Subminimum-Wages.php.

"*we now operate a program that gives permission*": Catherine Lhamon, interview with the author, 2020.

"*because a lack of belief in people with disabilities*": "Briefing: Subminimum Wages, People with Disabilities—Morning Session," YouTube Video, YouTube, https://www.youtube.com/watch?v=eXxUMNfH85A&feature=emb_title.

This is something the generations that came up: Julie Christensen, interview with the author, 2020.

76 "*Inclusion measures after they're through the door*": Marcelle Ciampi, interview with the author, 2019.

"*crashing and burning*": John Marble, interview with the author, 2019.

77 "*If we can address all those issues*": Marcelle Ciampi, interview with the author, 2019.

wore a shirt supporting neurodiversity to work: Chris Ereneta, Chris Williams, interview with the author, 2020.

Square has several employee resource groups (ERGs): "ERGs, COVID-19, and Community, Inclusion & Diversity at Square," Square, 2020, https://squareup.com/us/en/l/diversity/ergs-community.

Ereneta, Williams, and a third colleague: Chris Ereneta, interview with the author, 2020.

"*We have a lot of activity around attention differences*": Chris Ereneta, interview with the author, 2020.

78 *Claudia Ng, who considers herself neurodivergent:* Claudia Ng, e-mail with the author, 2020.
she long suspected she might have some attention issues: Claudia Ng, interview with the author, 2020.
starting with his time before working in finance: Chris Williams, interview with the author, 2019.
79 *recruit neurodivergent people:* Tyneisha Harris, interview with the author, 2020.
Ereneta said he wants to make sure Square: Chris Ereneta, interview with the author, 2020.
80 *Pivot Diversity:* John Marble, interview with the author, 2019.
81 *he likens being autistic to being French:* John Marble, interview with the author, 2018.
if you put autistic people in a room: John Marble, interview with the author, 2019.

4. "Gimme Shelter"

83 *Julia Bascom has advised presidential candidates:* Julia Bascom, interview with the author, 2020.
an autistic person with limited speaking capacity: Silberman, *NeuroTribes,* 44–46.
continues to live at home and requires 24/7 support: Reporting by the author, 2020.
her support person, Colton Callahan: Julia Bascom and Colton Callahan, interview with the author, 2020.
84 *"One is entitled to wonder":* Pitney, *The Politics of Autism,* 15.
"chronic undifferentiated schizophrenia": Jessica Wright, "The Missing Generation," *Spectrum,* December 9, 2015, https://www.spectrumnews.org/features/deep-dive/the-missing-generation/.
their parents sent them to institutions: Silberman, *NeuroTribes,* 279.
85 *Bettelheim subjected his children to strikes and whips with a belt:* Silberman, *NeuroTribes,* 206–7.
compared him to a cult leader: Richard Bernstein, "Accusations of Abuse Haunt the Legacy of Dr. Bruno Bettelheim," *New York Times,* November 4, 1990, https://www.nytimes.com/1990/11/04/weekinreview/ideas-trends-accusations-of-abuse-haunt-the-legacy-of-dr-bruno-bettelheim.html.
used corporal punishment: Ronald Angres, "Who, Really, Was Bruno Bettelheim?" *Commentary,* October 1, 1990, https://www.commentarymagazine.com/articles/ronald-angres/who-really-was-bruno-bettelheim/.
the realities of these institutions: Cal Montgomery, interview with the author, 2019.
Montgomery was institutionalized: Cal Montgomery, correspondence with the author, 2020.
"And I cheerfully signed myself in voluntarily": Cal Montgomery, interview with the author, 2019.
Montgomery spent three short-term stays: Cal Montgomery, correspondence with the author, 2020.
86 *"This second institution [in Massachusetts]":* Cal Montgomery, interview with the author, 2019.
the belief that autism is fundamentally different: Samantha Crane, e-mail correspondence with the author, 2019.
Joseph Sullivan: Silberman, *NeuroTribes,* 373–75.
87 *It was likely that they were able to develop the very skills:* Silberman, *NeuroTribes.*

This proved to be a step too far for autism "experts": Silberman, *NeuroTribes,* 375.
"the 'happy ending' in the original script is simply not realistic": Silberman, *NeuroTribes,* 376.
"I felt betrayed politically": Silberman, *NeuroTribes.*
"They were in institutions, basically": Steve Silberman, interview with the author, 2020.
even though Rimland did not institutionalize: Silberman, *NeuroTribes,* 376.
"zealous advocates for their own": Bernard Rimland, "Beware the Advozealots: Mindless Good Intentions Injure the Handicapped," *Autism Research Review International* 7, no. 4 (1993), https://web.archive.org/web/20190708132949/http://ariconference.com/ari/newsletter/074/page3.pdf.

88 *Leo Rosa seemed to be a generally happy person:* Shannon des Roches Rosa, interview with the author, 2020.

89 *support needs as "very middle of the road":* Julia Bascom, interview with the author, 2020.

90 *"Julia was looking for a roommate–slash–support person":* Colton Callahan, interview with the author, 2020.
"I could probably prepare two very simple meals": Julia Bascom, interview with the author, 2020.

91 *"an individual, a couple or a family":* Arizona Developmental Disabilities Planning Council, "Shared Living," https://www.nasddds.org/uploads/documents/Shared_Living_Brief.pdf.
found that 5 percent of people with intellectual or developmental disabilities: S. A. Larson et al., "In-Home and Residential Long-Term Supports and Services for Persons with Intellectual or Developmental Disabilities: Status and Trends through 2017," University of Minnesota, 2020, https://Ici-s.Umn.Edu/Files/ACHyYaFjMi/Risp_2017. Minneapolis.
grocery shopping is not something she can do: Julia Bascom and Colton Callahan, interview with the author, 2020.

92 *"And she's made that connection in her head":* Julia Bascom and Colton Callahan, interview with the author, 2020.

93 *Congress enacted 1915(c) of the Social Security Act:* Mary Jean Duckett and Mary R. Guy, "Home- and Community-Based Services Waivers," *Health Financing Review* 22, no. 1 (2000): 123–25, https://www.cms.gov/Research-Statistics-Data-and-Systems/Research/HealthCareFinancingReview/Downloads/00Fallpg123.pdf.

94 *the Americans with Disabilities Act "instructed states":* Pitney, *The Politics of Autism,* 92.
"First, institutional placement of persons": Olmstead v. L.C., U.S. 853 (1975).
70 percent of spending went to HCBS waivers: Pitney, *The Politics of Autism,* 92.

95 *"They're based off of people with intellectual disabilities":* Julia Bascom, interview with the author, 2020.
the noise could cause sensory overload: Cal Montgomery, interview with the author, 2020.
But states also have the capacity to cap HCBS waivers: MaryBeth Musumeci, Priya Chidambaram, and Molly O'Malley Watts, "Key Questions About Medicaid Home- and Community-Based Services Waiver Waiting Lists," Kaiser Family Foundation, 2019, https://www.kff.org/medicaid/issue-brief/key-questions-about-medicaid-home-and-community-based-services-waiver-waiting-lists/.
a backlog of 3,791 people: Amanda Seitz, "State Hopes to Fix Disability Agency's Wait-

list," WCPO, June 22, 2017, https://www.wcpo.com/news/insider/medicaid-waiver-wait-grows-for-hamilton-county-developmental-disability-services-agency.

five thousand individuals over a six-year period: Dan McKay, "New Mexico to Redesign Major Disability Program," *Las Cruces Sun-News,* July 10, 2019, https://www.lcsun-news.com/story/news/local/new-mexico/2019/07/10/new-mexico-disability-program-gov-lujan-grisham-healthcare-insurance/1693271001/.

707,000 people on waiting lists: Musumeci et al., "Key Questions About Medicaid."

96 *One study published in* Health Services Research *in 2019:* Michelle LaClair et al., "The Effect of Medicaid Waivers on Ameliorating Racial/Ethnic Disparities Among Children with Autism," *Health Services Research* 54, no. 4 (May 27, 2019), https://doi.org/10.1111/1475-6773.13176.

Maryland, North Dakota, Arkansas, Montana: Doug Leslie, correspondence with the author, 2020.

the waivers improved their overall family quality of life: Karen Goldrich Eskow and Jean Ann Summers, "Family Perceptions of the Impacts of a Home- and Community-Based Services Autism Waiver: Making Family Life Possible," *Journal of Applied Research in Intellectual Disabilities* 32, no. 1 (August 27, 2018), https://doi.org/10.1111/jar.12518.

found that HCBS waivers: Douglas L. Leslie et al., "The Effects of Medicaid Home- and Community-Based Services Waivers on Unmet Needs Among Children With Autism Spectrum Disorder," *Medical Care* 55, no. 1, (January 2017): 57–63), https://doi.org/10.1097/mlr.0000000000000621, 159–71.

for someone in an institution was $188,318: "Preserving Our Freedom Ending Institutionalization of People with Disabilities During and After Disasters," National Council on Disability, May 24, 2019, https://ncd.gov/sites/default/files/NCD_Preserving_Our_Freedom_508.pdf.

97 "*Even when a person*": Samantha Crane, correspondence with the author, 2019.

Medicaid specifically prohibits paying for room and board: Janet O'Keeffe et al., "Understanding Medicaid Home and Community Services: A Primer 2010 Edition," U.S. Department of Health and Human Services, 2010, https://aspe.hhs.gov/system/files/pdf/76201/primer10.pdf.

"*they often can charge lower room and board fees*": Samantha Crane, correspondence with the author, 2019.

most HCBS coverage is optional: "Medicaid Long-Term Services and Supports: An Overview of Funding Authorities," Kaiser Family Foundation, September 25, 2013, https://www.kff.org/medicaid/fact-sheet/medicaid-long-term-services-and-supports-an-overview-of-funding-authorities/.

states spent anywhere between 21 to 78 percent of the money: Erica L. Reaves and MaryBeth Musumeci, "Medicaid and Long-Term Services and Supports: A Primer," Kaiser Family Foundation, December 15, 2015, https://www.kff.org/medicaid/report/medicaid-and-long-term-services-and-supports-a-primer/view/footnotes/#footnote-172646-27.

"*Because group homes and institutions serve multiple people at once*": Reaves and Musumeci, "Medicaid and Long-Term Services."

98 *could use Medicaid dollars to pay for HCBS:* Centers for Medicare and Medicaid Services, 2014, "Medicaid Program; State Plan Home- and Community-Based Services, 5-Year Period for Waivers, Provider Payment Reassignment, and Home- and Commu-

nity-Based Setting Requirements for Community First Choice and Home and Community-Based Services (HCBS) Waivers," https://www.federalregister.gov/documents/2014/01/16/2014-00487/medicaid-program-state-plan-home-and-community-based-services-5-year-period-for-waivers-provider.

requirements for home- and community-based settings: "Home- and Community-Based Settings Fact Sheet," *Arc,* 2016, https://thearc.org/wp-content/uploads/forchapters/2016%20Home%20%26%20Community%20Based%20Settings%20Fact%20Sheet.pdf.

outlined four types of group settings: "Guidance on Settings That Have the Effect of Isolating Individuals Receiving HCBS from the Broader Community," Centers for Medicare and Medicaid Services, https://www.medicaid.gov/medicaid/hcbs/downloads/settings-that-isolate.pdf.

when the policy was to "protect" them: Deborah S. Metzel and Pamela M. Walker, "The Illusion of Inclusion: Geographies of the Lives of People with Developmental Disabilities in the United States," *Disability Studies Quarterly* 21, no. 4 (Fall 2001).

future setting for his own son Mark: Autistic Adults in Bittersweet Farms, ed. Norman S. Giddan and Jane J. Giddan (Binghamton, NY: Haworth Press, 1991), 1.

"*Often these farm arrangements are custodial*": Kit Mead, "Stop Isolating Autistic Adults and Calling It 'Community-Based Housing,'" *Thinking Person's Guide to Autism,* August 3, 2016, http://www.thinkingautismguide.com/2016/08/stop-isolating-autistic-adults-and.html.

99 "*primarily of people with disabilities*": "Guidance on Settings," Centers for Medicare and Medicaid Services.

did not track abuse claims: Danny Hakim, "State Faults Care for the Disabled," *New York Times,* March 22, 2012, https://www.nytimes.com/2012/03/22/nyregion/new-york-state-draft-report-finds-needless-risk-in-care-for-the-disabled.html.

use of restraints: Annie Waldman, "Kids Get Hurt at Residential Schools While States Look On," ProPublica, December 15, 2015, https://www.propublica.org/article/kids-get-hurt-at-residential-schools-while-states-look-on.

operated and controlled by the same group: "Guidance on Settings," Centers for Medicare and Medicaid Services.

"*They finally get their alleged HCBS*": Julia Bascom, interview with the author, 2020.

100 *National Council on Severe Autism:* Amy Lutz, "National Council on Severe Autism (NCSA) Launches," *Psychology Today,* January 14, 2019, https://www.psychologytoday.com/gb/blog/inspectrum/201901/national-council-severe-autism-ncsa-launches.

"*But others—like Jonah*": Amy S. F. Lutz, "Adults with Disabilities Deserve Right to Choose Where to Live," *Spectrum News,* May 2, 2017, https://www.spectrumnews.org/opinion/viewpoint/adults-disabilities-deserve-right-choose-live/.

"*launched a strike on intentional communities*": Paul Solataroff, "Luke's Best Chance: One Man's Fight for His Autistic Son," *Rolling Stone,* July 27, 2016, https://www.rollingstone.com/culture/culture-features/lukes-best-chance-one-mans-fight-for-his-autistic-son-93049/.

a founding member of NCSA: Lutz, "National Council on Severe Autism."

101 *gave states until 2022:* "CMS Announces Extension for States Under Medicaid Home and Community-Based Settings Criteria," Centers for Medicare and Medicaid Services, May 9, 2017, https://www.cms.gov/newsroom/press-releases/cms-announces-extension-states-under-medicaid-home-and-community-based-settings-criteria.

March 17, 2023: Calder A. Lynch, "Re: Home- and Community-Based Settings Reg-

ulation—Implementation Timeline Extension and Revised Frequently Asked Questions," Centers for Medicare and Medicaid Services, July 14, 2020.
released new guidance on the settings rule: "CMS Issues New Guidance on State Implementation of Home- and Community-Based Services Regulation," Centers for Medicare and Medicaid Services, March 22, 2019, https://www.cms.gov/newsroom/press-releases/cms-issues-new-guidance-state-implementation-home-and-community-based-services-regulation.
"not automatically presumed to have qualities of an institution": Chris Traylor, "Frequently Asked Questions: HCBS Settings Regulation Implementation Heightened Scrutiny Reviews of Presumptively Institutional Settings," Centers for Medicare and Medicaid Services, March 22, 2019, https://www.medicaid.gov/federal-policy-guidance/downloads/smd19001.pdf.
"often lack the flexibility needed": "CMS Issues New Guidance," Centers for Medicare and Medicaid Services.
calling it a victory for "choice": "A Federal Policy Victory for Adults with Autism," National Council on Severe Autism, March 25, 2019, https://www.ncsautism.org/blog/2019/3/24/qkyj2l69k3hgx9ex4xrydv8zo7dz3w.

102 *which supporters describe:* "Intentional or 'Campus' Communities for Adults with Autism and Developmental Disabilities," Autism Society San Francisco Bay Area, 2020, https://www.sfautismsociety.org/list-of-intentional-communities.html.
nothing like the institutions of old: "Full Committee Meeting," Interagency Autism Coordinating Committee, January 16, 2019, https://iacc.hhs.gov/meetings/iacc-meetings/2019/full-committee-meeting/january16/transcript_011619.pdf.
there is no data proving: Lutz, "Adults with Disabilities."
"I think if someone thinks an intentional community": Julia Bascom, interview with the author, 2020.
"designed so that people have no need to leave the campus": Shannon Des Roches Rosa, "Autism and Adult Housing Choices: Separating Myths from Facts," *Thinking Person's Guide to Autism*, August 22, 2016, http://www.thinkingautismguide.com/2016/08/autism-and-adult-housing-choices.html.

103 *"the same old people are working it":* Cal Montgomery, interview with the author, 2020.
the Rosa family's insurance covers in-home services: Shannon Des Roches Rosa, interview with the author, 2020.

104 *"Guardianship involves substituted decision-making":* Michael Kendrick, interview with the author, 2019.

105 *deeply Republican Texas in 2015:* "Supported Decision-Making Overview," Disability Rights Texas, February 12, 2020.
followed by Delaware: "Supported Decision Making in Delaware," Delaware Department of Social Services, 2018, https://www.dhss.delaware.gov/dhss/dsaapd/files/supported_decision_making_information.pdf.
states that have laws on their books include: Zachary Allen and Dari Pogach, "More States Pass Supported Decision-Making Agreement Laws," American Bar Association, October 1, 2019, https://www.americanbar.org/groups/law_aging/publications/bifocal/vol-41/volume-41-issue-1/where-states-stand-on-supported-decision-making/.
Samantha Crane: Julia Bascom, interview with the author, 2020.
They weren't able to get that bigger bill passed: Anna Krieger, interview with the author, 2019.
Money Follows the Person: "Money Follows the Person," Centers for Medicare and

Medicaid Services, 2020, https://www.medicaid.gov/medicaid/long-term-services-supports/money-follows-person/index.html.

106 *which has caused a drop:* H. Stephen Kaye and Joe Caldwell, "Short-Term Money Follows the Person Extensions Resulted in a Significant Drop in State Efforts to Transition People Out of Institutions," Community Living Policy Center, Brandeis University, May 29, 2020, http://www.advancingstates.org/hcbs/article/short-term-money-follows-person-extensions-resulted-significant-drop-state-efforts.

"*If you don't have funding available to you*": Samantha Crane, correspondence with the author, 2019.

108 "*the risk of failure*": Ruti Regan, "Uncertain Abilities and the Right to Fail," January 18, 2016, https://www.realsocialskills.org/blog/uncertain-abilities-and-the-right-to-fail.

"*I get to make dumb choices and have no one check them*": Colton Callahan, interview with the author, 2020.

109 *I was really glad that I was able to do it:* Julia Bascom, interview with the author, 2020.

"*it's hard to balance the consequences of talking about support needs*": Julia Bascom, interview with the author, 2020.

110 "*looks like someone who's in charge of their life*": Julia Bascom, interview with the author, 2020.

5. "Somebody Get Me a Doctor"

111 *the so-called medical model:* Sara Goering, "Rethinking Disability: The Social Model of Disability and Chronic Disease," *Current Review of Musculoskeletal Medicine* 8, no. 2 (June 2015), https://www.ncbi.nlm.nih.gov/pmc/articles/PMC4596173/#CR2.

society is not accessible to disabled people: Aiyanna Bailin, "Clearing Up Some Misconceptions About Neurodiversity," *Scientific American,* June 6, 2019, https://blogs.scientificamerican.com/observations/clearing-up-some-misconceptions-about-neurodiversity/.

112 "*people whose brains work in atypical ways*": Bailin, "Clearing Up Some Misconceptions."

"*Nearly all parents' hands went up for services*": Silberman, *NeuroTribes,* 304.

"*When you look at the tremendous increase*": Michelle Ye Hee Lee, "Trump's Claim That There's 'Tremendous Amount of Increase' in Autism Cases," *Washington Post,* February 16, 2017, https://www.washingtonpost.com/news/fact-checker/wp/2017/02/16/trumps-claim-that-theres-tremendous-amount-of-increase-in-autism-cases/.

writing a profile for Roll Call: Eric Garcia, "Booker Tacks Toward Democratic Party's Base," *Roll Call,* September 15, 2017, https://www.rollcall.com/2017/09/15/booker-tacks-toward-democratic-partys-base/.

113 *like pissing down your leg:* CQ *Roll Call* staff, "When the President of the United States Talks," *Roll Call,* March 14, 2013, https://www.rollcall.com/2013/03/14/when-the-president-of-the-united-states-talks/.

"*Cancer rates, autism rates, asthma rates*": Congressman John Conyers Jr., "Will President Trump Undo a Signature Civil Rights Movement Accomplishment," Facebook, September 12, 2017, https://www.facebook.com/206947066849/videos/10154763272686850/.

"*unreasonable, irrational, impractical love*": Franklin Foer, "Cory Booker's Theory of Love," *Atlantic,* December 17, 2018, https://www.theatlantic.com/politics/archive/2018/12/cory-booker-talks-about-love-and-eyes-presidential-run/578242/.

a Stanford graduate and Rhodes Scholar: Janell Ross, "Six Noteworthy Things About Cory Booker," *Washington Post,* July 25, 2016, https://www.washingtonpost.com/news/the-fix/wp/2016/07/25/six-noteworthy-things-about-cory-booker/.

114 *the drug metrazol:* Silberman, *NeuroTribes,* 209.

psychotropic drugs like LSD: Jeff Sigafoos et al., "Flashback to the 1960s: LSD in the Treatment of Autism," *Developmental Neurorehabilitation* 10, no. 1 (January–March 2007), https://pubmed.ncbi.nlm.nih.gov/17608329/.

was "thawing" them out: Grinker, *Unstrange Minds,* 80.

claimed that 92 percent of speaking children: Silberman, *NeuroTribes,* 207–8.

that more people would be diagnosed: Gernsbacher, Dawson, Goldsmith, "Three Reasons Not to Believe in an Autism Epidemic."

officially removed the word cure*:* Dahl, "A Leading Autism Organization."

Bernard Rimland's Autism Research Institute: Jeffrey Baker, "Mercury, Vaccines, and Autism: One Controversy, Three Histories," *American Journal of Public Health* 98, no. 2 (2007), https://ajph.aphapublications.org/doi/10.2105/AJPH.2007.113159.

Jonathan Shestack and Portia Iversen: Portia Iversen, "Autism Research and Advocacy Related Biosketch," October 2018, https://portiaiversen.com/autism/.

115 "*it is up to parents to lead the revolution*"*:* Bernard Rimland, "Fighting for Our Children: Advocacy in the Age of Environmental Awareness," *Autism Advocate* 45 (2006), https://www.as-az.org/wp-content/uploads/2020/05/environmental-health-issue.pdf.

"*leaky gut syndrome*"*:* Linda Steele, "'It Is Not About the Science. It's About Belief,'" *Guardian,* December 5, 2001, https://www.theguardian.com/education/2001/dec/05/medicalscience.familyandrelationships.

vaccine particles prevented the breakdown of certain foods: Steven Novella, "The Anti-Vaccination Movement," *Skeptical Inquirer* (November/December 2007).

Rimland warned against gluten and casein: Silberman, *Neurotribes,* 334.

sixty-six autistic children: Anna Piwowarczyk et al., "Gluten-Free Diet in Children with Autism Spectrum Disorders: A Randomized, Controlled, Single-Blinded Trial," *Journal of Autism and Developmental Disorders* 50 (October 2019): 482–90, https://link.springer.com/article/10.1007%2Fs10803-019-04266-9.

116 *named the compound the Miracle Mineral Solution:* Brandy Zadrozny, "Parents Are Poisoning Their Children with Bleach to 'Cure' Autism," *NBC News,* May 21, 2019, https://www.nbcnews.com/tech/internet/moms-go-undercover-fight-fake-autism-cures-private-facebook-groups-n1007871.

"*false—and dangerous—claims*"*:* "Danger: Don't Drink Miracle Mineral Solution or Similar Products," U.S. Food and Drug Administration, August 12, 2019, https://www.fda.gov/consumers/consumer-updates/danger-dont-drink-miracle-mineral-solution-or-similar-products.

placing kids in hyperbaric chambers: Trine Tsouderos and Patricia Callahan, "On Shaky Ground with Alternative Treatments to Autism," *Los Angeles Times,* December 7, 2009, https://www.latimes.com/archives/la-xpm-2009-dec-07-la-hew-autism-day-two7-2009dec07-story.html.

chelation therapy: Jay L. Hoecker, "Autism Treatment: Can Chelation Therapy Help?" Mayo Clinic, November 23, 2016, https://www.mayoclinic.org/diseases-conditions/autism-spectrum-disorder/expert-answers/autism-treatment/faq-20057933.

meant to help people improve basic skills: "Applied Behavior Analysis," *Psychology Today,* 2020, https://www.psychologytoday.com/us/therapy-types/applied-behavior-analysis.

using conditioning: Silberman, *NeuroTribes,* 286.

117 *breaking down bigger skills:* Pitney, *The Politics of Autism,* 26.

rewarding good behavior with food: Don Moser, "Screams, Slaps and Love," *Life,* 1965, http://neurodiversity.com/library_screams_1965.pdf.

"*rock themselves back and forth*": Paul Chance, "O. Ivar Lovaas Interview with Paul Chance," *Psychology Today,* January 1974, http://neurodiversity.com/library_chance_1974.pdf.

Lovaas removed the punitive aversive methods: Pitney, *The Politics of Autism,* 20.

"*gold standard*": Pitney, *The Politics of Autism*, 57.

insurers to cover ABA: "Autism and Insurance Coverage/State Laws," National Conference of State Legislatures, August 8, 2018, https://www.ncsl.org/research/health/autism-and-insurance-coverage-state-laws.aspx.

he went through ABA: Hari Srinivasan, correspondence with the author, 2020.

difficulty with sensory processing: Lisa Jo Rudy, "Autism and Sensory Overload," *Verywell Health,* November 25, 2019, https://www.verywellhealth.com/autism-and-sensory-overload-259892.

"*It was groundhog day as I was stuck in the same lessons*": Hari Srinivasan, correspondence with the author, 2020.

118 *was inspired to pursue a master's degree in ABA:* Amy Gravino, interview with the author, 2020.

"*I want to be cured*": Alisa Opar, "In Search of a Truce in the Autism Wars," *Spectrum News,* April 24, 2019, https://www.spectrumnews.org/features/deep-dive/search-truce-autism-wars/.

119 "*severe neurodevelopmental disability*": Moheb Costandi, "Against Neurodiversity," *Aeon,* September 12, 2019, https://aeon.co/essays/why-the-neurodiversity-movement-has-become-harmful.

neurodivergent people deserve the same rights as others: Sara Luterman, "Autistic Advocates Clash with Autism Parents at Government Committee Meeting," *NOS*, October 27, 2017, http://nosmag.org/parents-and-autistic-adults-clash-at-autism-committee-meeting-iacc-interagency-autism-coordinating-committee/.

some autistic people have more impairments: Bailin, "Clearing Up Some Misconceptions."

gastrointestinal issues: Katherine Zeratsky, "Autism Spectrum Disorder and Digestive Symptoms," Mayo Clinic, May 21, 2019, https://www.mayoclinic.org/diseases-conditions/autism-spectrum-disorder/expert-answers/autism-and-digestive-symptoms/faq-20322778.

"*considered worthy individuals in and of themselves*": Nancy Bagatell, "From Cure to Community: Transforming Notions of Autism," *Journal for the Society for Psychological Anthropology,* March 11, 2010, https://anthrosource.onlinelibrary.wiley.com/doi/epdf/10.1111/j.1548-1352.2009.01080.x.

120 *"Our duty in autism is not to cure"*: Garcia, "I'm Not Broken."

she had moved back in with her parents: Lydia Wayman, interview with the author, 2020.

121 *A separate nurse:* Lydia Wayman, interview with the author, 2020.

The way Lydia's care is paid for: Lydia Wayman, e-mail to the author, 2020.

"*I guess it started with my stomach*": Lydia Wayman, interview with the author, 2020.

four times between January and April of that year: Lydia Wayman, interview with the author, 2019.

her blood vessels stay dilated: Lydia Wayman, interview with the author, 2020.

122 *her illness prevented her from participating:* Lydia Wayman, correspondence with the author, 2020.

"*I just didn't always agree with how they got implemented*": Lydia Wayman, correspondence with the author, 2020.

"primary" diagnosis was autism: Lydia Wayman, interview with the author, 2019.

"*isn't supposed to exist*": Lydia Wayman, interview with the author, 2019.

in a nursing home for fifteen months: Lydia Wayman, interview with the author, 2020.

mental illness or intellectual disabilities: Lydia Wayman, e-mail to the author, 2020.

"*They were one of the few*": Nancy Wayman, interview with the author, 2020.

123 *prior to Lydia's time at the nursing home*: Lydia Wayman, e-mail to the author, 2020.

"*I just didn't know that you had to play social games*": Lydia Wayman, interview with the author, 2020.

"*she was with me through the whole entire process*": Lydia Wayman, interview with the author, 2019.

"*but I didn't realize I was giving her that impression*": Lydia Wayman, interview with the author, 2020.

she wasn't faking anything: Lydia Wayman, e-mail to the author, 2020.

124 "*you're so lost, and you've got to process it*": Lydia Wayman, interview with the author, 2020.

bad or mean doctors: Lydia Wayman, e-mail to the author, 2020.

everything began to make sense: Lydia Wayman, e-mail to the author, 2020.

125 *Lydia can just look at her mom:* Lydia Wayman, interview with the author, 2020.

"vomiting, chest pains, and a persistent cough": Organ Transplant Discrimination Against People with Disabilities, National Council on Disability, September 25, 2019, https://ncd.gov/sites/default/files/NCD_Organ_Transplant_508.pdf.

"psychiatric issues, autism": Lenny Bernstein, "People with Autism, Intellectual Disabilities Fight Bias in Transplants," *Washington Post,* March 4, 2017, https://www.washingtonpost.com/national/health-science/people-with-autism-intellectual-disabilities-fight-bias-in-transplants/2017/03/04/756ff5b8-feb2-11e6-8f41-ea6ed597e4ca_story.html.

though his condition is stable for now: Karen Corby, correspondence with the author, 2020.

her emergency department care improved after she contacted: Lydia Wayman, interview with the author, 2020.

126 "*a doctor, a resident, and three interns*": Lydia Wayman, correspondence with the author, 2020.

"*purposely designing a place for people with ASD*": Arvind Venkat, interview with the author, 2020.

127 "*We're not teaching medical students*": Lydia Wayman, interview with the author, 2020.

estimated that 2.2 percent of adults were autistic: Fox, "First US Study of Autism."

that 20 to 30 percent of all autistic individuals: "Autism Spectrum Disorder Fact Sheet," National Institute of Neurological Disorders and Stroke, 3–13, 2020, https://www.ninds.nih.gov/Disorders/Patient-Caregiver-Education/Fact-Sheets/Autism-Spectrum-Disorder-Fact-Sheet.

128 *sudden unexpected death in people with epilepsy:* "SUDEP," Epilepsy Foundation, 2020, https://www.epilepsy.com/learn/early-death-and-sudep/sudep.

surveyed more than 27,000 autistic adults: Tatja Hirvikoski et al., "Premature

Mortality in Autism Spectrum Disorder," *British Journal of Psychiatry* 208, no. 3 (March 2016), https://www.cambridge.org/core/journals/the-british-journal-of-psychiatry/article/premature-mortality-in-autism-spectrum-disorder/4C9260DB64DFC29AF945D32D1C15E8F2.

followed by suicide: Ann Griswold, "Large Swedish Study Ties Autism to Early Death," *Spectrum News,* December 11, 2015, https://www.spectrumnews.org/news/large-swedish-study-ties-autism-to-early-death/.

only 2 percent: Office of Autism Research Coordination, National Institute of Mental Health, Autistica, Canadian Institutes of Health Research, and Macquarie University, *2016 International Autism Spectrum Disorder Research Portfolio Analysis Report,* 41.

environment in autism's development: Office of Autism Research Coordination, *2016 International Autism Spectrum Disorder,* 25.

more than half of the money spent on autism: Office of Autism Research Coordination, *2016 International Autism Spectrum Disorder,* 41.

129 "*the clarity and precision of the language*": Christina Nicolaidis et al., "Comparison of Healthcare Experiences in Autistic and Non-Autistic Adults: A Cross-Sectional Online Survey Facilitated by an Academic-Community Partnership," *Journal of General Internal Medicine* 28, no. 6 (June 6, 2013), https://www.ncbi.nlm.nih.gov/pmc/articles/PMC3663938/.

"*more likely to experience depression*": Chloe C. Hudson, Layla Hall, and Kate L. Harkness, "Prevalence of Depressive Disorders in Individuals with Autism Spectrum Disorder: A Meta-Analysis," *Journal of Abnormal Child Psychology* 47, no. 1 (March 1, 2018), https://link.springer.com/article/10.1007%2Fs10802-018-0402-1.

130 *Cal Montgomery noted:* Cal Montgomery, interview with the author, 2019.

6. "Ain't Talkin' 'Bout Love"

132 *some autistic kids didn't speak:* Chris Williams, interview with the author, 2020.

"*just feeling very happy to be in her presence*": Chris Williams, interview with the author, 2020.

133 "'*I'm going to marry that boy*'": Cori Williams, interview with the author, 2020.

134 "*tend to be a bit naive and trusting*": Matthew Roth and Jennifer Gillis, "'Convenience with the Click of a Mouse': A Survey of Adults with Autism Spectrum Disorder on Online Dating," *Sexuality and Disability* 33, no. 1 (December 31, 2014), https://link.springer.com/article/10.1007%2Fs11195-014-9392-2.

when Cassidy Luna was diagnosed: Chris Williams, interview with the author, 2019.

"*Holy crap, I might also be as well*": Cori Williams, interview with the author, 2020.

135 "*The desire, the drive and the social knowledge*": Suzanne Leigh, "A Long Shadow Is Lifted on Asperger's in Adults," *USA Today,* July 23, 2007, https://usatoday30.usatoday.com/news/health/2007-07-23-adult-diagnosis_N.htm.

neurotypical people asking their autistic classmates to prom: Alexandra Zaslow, "Cheerleader Asks Special Needs Student to Prom in Sweetest Way Possible," *Today,* February 29, 2016, https://www.today.com/health/cheerleader-asks-special-needs-student-prom-sweetest-way-possible-t76571.

"*when you have it by yourself*": Cori Williams, interview with the author, 2020.

136 *how women could navigate "high-tech relationships"*: Silberman, *NeuroTribes,* 252.

"*barely changed another word in the text*": Silberman, *NeuroTribes,* 252.

140 *"It felt like I had to mask more":* Chris Williams, interview with the author, 2020.
141 *"It'd be interesting to ponder that":* Chris and Cori Williams, interview with the author, 2020.
"It was not tears of unhappiness": Anlor Davin, interview with the author, 2020.
a nonprofit that focuses on supporting autistic people: Greg Yates, interview with the author, 2020.
142 *"like the sensory stuff":* Anlor Davin, interview with the author, 2020.
denounced by the United Nations: Angelique Chrisafis, "France Unveils €340m Plan to Improve Rights of People with Autism," *Guardian,* April 6, 2018, https://www.theguardian.com/world/2018/apr/06/france-to-unveil-340m-plan-to-improve-rights-of-people-with-autism.
send their children to institutions: Angelique Chrisafis, "'France Is 50 Years Behind': The 'State Scandal' of French Autism Treatment," *Guardian,* February 8, 2018, https://www.theguardian.com/world/2018/feb/08/france-is-50-years-behind-the-state-scandal-of-french-autism-treatment.
"Our autisms are not the same but there is some overlap": Greg Yates, interview with the author, 2020.
143 *Greg slammed doors:* Anlor Davin, interview with the author, 2020.
because of the autism: Greg Yates, interview with the author, 2020.
"Fears and stereotypes often impede individuals with disabilities": "Human Sexuality and ASD," *Sun,* July–September 2017, https://www.delautism.org/wp-content/uploads/2017/10/The-Sun-July-Sept2017.pdf.
144 *"you will go to jail":* Amy Gravino, interview with the author, 2020.
did not know that their children had masturbated: J. Dewinter et al., "Parental Awareness of Sexual Experience in Adolescent Boys with Autism Spectrum Disorder," *Journal of Autism and Developmental Disorders* 46, no. 2 (October 25, 2019), https://www.ncbi.nlm.nih.gov/pmc/articles/PMC4724358/.
"you're thinking that they're innocent": Amy Gravino, interview with the author, 2020.
146 *have been featured on* Good Morning America: Thea Trachtenberg and Lindsay Goldwert, "Couple Lives With Autism, Comfort of Each Other," *Good Morning America,* February 24, 2009, https://abcnews.go.com/GMA/OnCall/story?id=6952013&page=1.
were the subjects of a documentary: Lisa Bonos, "Love on the Spectrum: How Autism Brought One Couple Together," *Washington Post,* January 8, 2016, https://www.washingtonpost.com/news/soloish/wp/2016/01/08/love-on-the-spectrum-how-autism-brought-one-couple-together/.
they were in their first year of marriage: Eric Garcia, "Autistic Men Don't Always Understand Consent. We Need to Teach Them," *Washington Post,* April 27, 2017, https://www.washingtonpost.com/posteverything/wp/2017/04/27/autistic-men-dont-always-understand-consent-we-need-to-teach-them/.
"We didn't have a lot of those conversations": Lindsey Nebeker, interview with the author, 2016.
147 *it's nice to have the consent to do it:* Dave Hamrick, interview with the author, 2016.
309 autistic individuals: R. George and M. A. Stokes, "Sexual Orientation in Autism Spectrum Disorder," *Autism Research* 11, no. 1 (2018), https://pubmed.ncbi.nlm.nih.gov/29159906/.
"That's a whole other kind of barrier": Amy Gravino, interview with the author, 2020.

148 *is bisexual and married to a man:* Jessica Benham, interview with the author, 2020.
is a lesbian: Haley Moss, "Julia Bascom on the Amazing, Vibrant and Resilient Autistic Community," *Geek Club Books,* March 28, 2019.
"a tap on the hand": John Marble, interview with the author, 2018.
they chalk it up to autism: Melanie Yergeau, interview with the author, 2018.
coauthored: George A. Rekers, O. Ivar Lovaas, and Benson Low, "The Behavioral Treatment of a 'Transsexual' Preadolescent Boy," *Journal of Abnormal Child Psychology* 2, no. 2 (June 1974), https://link.springer.com/article/10.1007%2FBF00919093; George Rekers and Ole Ivar Lovaas, "Behavioral Treatment of Deviant Sex-Role Behaviors in a Male Child," *Journal of Applied Behavior Analysis* 7, no. 2 (1974), https://www.ncbi.nlm.nih.gov/pmc/articles/PMC1311956/pdf/jaba00060-0003.pdf; G. A. Rekers et al., "Sex-Role Stereotypy and Professional Intervention for Childhood Gender Disturbance," *Professional Psychology* (1978): 127–36, https://content.apa.org/record/1980-23511-001; G. A. Rekers et al., "Child Gender Disturbances: A Clinical Rationale for Intervention," *Psychotherapy: Theory, Research, and Practice* (1977): 2–11, https://doi.apa.org/doiLanding?doi=10.1037%2Fh0087487.
conducted their work on a five-year-old boy: Silberman, *NeuroTribes,* 319–23.

149 *mother saw a television ad:* "The 'Sissy' Boy Experiment: A Documentary by CNN," YouTube/CNN/Savvas Tappi, June 13, 2011, https://www.youtube.com/watch?v=Q0lZBL2H4nI.

150 *"defrost enough to produce a child":* "Medicine: The Child Is Father," *Time,* July 25, 1960, http://content.time.com/time/subscriber/article/0,33009,826528-1,00.html.

151 *"Eleven Possibly Autistic Parents":* Sarah Deweerdt, "The Joys and Challenges of Being a Parent with Autism," *Atlantic,* May 18, 2017, https://www.theatlantic.com/family/archive/2017/05/autism-parenting/526989/.
the iPhone personal assistant Siri: Judith Newman, "To Siri, with Love," *New York Times,* October 17, 2014, https://www.nytimes.com/2014/10/19/fashion/how-apples-siri-became-one-autistic-boys-bff.html.
"I do not want Gus to have children": Judith Newman, *To Siri with Love: A Mother, Her Autistic Son, and the Kindness of Machines* (New York: HarperCollins, 2017), 115.

152 *"I cried myself to sleep":* Kaelan Rhywiol, "Why 'To Siri with Love' Is Incredibly Damaging to Autistic People Like Me," *Bustle,* December 8, 2017, https://www.bustle.com/p/why-i-believe-to-siri-with-love-by-judith-newman-is-a-book-that-does-incredible-damage-to-the-autistic-community-6780420.
the Iowa Supreme Court ruled: Grant Rodgers, "Ruling: Court OK Needed for Disabled Son's Vasectomy," *Des Moines Register,* April 18, 2014, https://www.usatoday.com/story/news/nation/2014/04/18/ruling-court-ok-needed-for-disabled-sons-vasectomy/7893917/.
"Hey, sweetie, Mom is right there": Chris Williams, interview with the author, 2020.

153 *"We don't have any idea what she's capable of yet":* Cori Williams, interview with the author, 2020.

155 *disagreed with each other during our interview:* Anlor Davin and Greg Yates, interview with the author, 2020.
that there are clashes: Chris and Cori Williams, interview with the author, 2020.

7. "Not Sure if You're a Boy or a Girl"

156 "*slower to get things*": Eryn Star, "Surviving Education Trauma: Teacher Abuse of Disabled Students," National Council on Independent Living, June 29, 2019, https://advocacymonitor.com/surviving-education-trauma-teacher-abuse-of-disabled-students/.

157 "*complicated my relationship with my gender identity*": Eryn Star, interview with the author, 2019.

158 *animal scientist Temple Grandin published:* Pripas-Kapit, "Historicizing Jim Sinclair's 'Don't Mourn for Us.'"

compared to just 6.6 per every 1,000 girls: Jon Baio et al., "Prevalence of Autism Spectrum Disorder Among Children Aged 8 Years—Autism and Developmental Disabilities Monitoring Network, 11 Sites, United States, 2014," Centers for Disease Control and Prevention, April 27, 2018, https://www.cdc.gov/mmwr/volumes/67/ss/ss6706a1.htm#suggestedcitation.

three girls out of eleven: Kanner, "Autistic Disturbances of Affective Contact."

"*an extreme variant of male intelligence*": Hans Asperger, "'Autistic Psychopathy' in Childhood," in *Autism and Asperger Syndrome,* ed. Uta Frith (Cambridge: Cambridge University Press, 1991), 37–92, https://blogs.uoregon.edu/autismhistoryproject/archive/hans-asperger-autistic-psychopathy-in-childhood-1944/.

"*autism can be considered as an extreme of the normal male profile*": Simon Baron-Cohen, "The Extreme Male Brain Theory of Autism," *Trends in Cognitive Science* 2, no. 2 (June 2002), https://pubmed.ncbi.nlm.nih.gov/12039606/.

159 *also argued that exposure to testosterone:* Emily Underwood, "Study Challenges Idea That Autism Is Caused by an Overly Masculine Brain," *Science,* September 3, 2019, https://www.sciencemag.org/news/2019/09/study-challenges-idea-autism-caused-overly-masculine-brain.

"*living proof that we're definitely not extreme males*": Lucie Kanfiszer, Fran Davies, and Suzanne Collins, "'I Was Just So Different': The Experiences of Women Diagnosed with an Autism Spectrum Disorder in Adulthood in Relation to Gender and Social Relationships," *Autism,* March 22, 2017, https://journals.sagepub.com/doi/abs/10.1177/1362361316687987.

extreme male brain theory: M. Remi Yergeau, *Authoring Autism* (Durham, NC: Duke University Press, 2018).

and how this plays a role in furthering the diagnosis gap: M. Remi Yergeau, interview with the author, 2018.

that this might be because girls' special interests: "Autism Signs Can Be Different in Girls," Cleveland Clinic, September 3, 2019, https://health.clevelandclinic.org/for-females-with-autism-differences-matter/.

more age-appropriate and less eccentric: Nancy Volkers, "Invisible Girls," *The Asha Leader,* April 1, 2018, https://leader.pubs.asha.org/doi/10.1044/leader.FTR1.23042018.48.

"*'that's not a special interest'*": Jess Commons, "'You're Just Kooky': Why Women with Autism Aren't Taken Seriously," Refinery29, September 1, 2019, https://www.refinery29.com/en-gb/autism-young-women.

than boys with ASD: Spencer C. Evans et al., "Sex/Gender Differences in Screening for Autism Spectrum Disorder: Implications for Evidence-Based Assessment," *Journal of*

Clinical Child and Adolescent Psychology 48, no. 6 (March 30, 2018), https://doi.org/10.1080/15374416.2018.1437734.

160 "*over the diagnostic threshold*": Katharina Dworzynski et al., "How Different Are Girls and Boys Above and Below the Diagnostic Threshold for Autism Spectrum Disorders?," *Journal of the American Academy of Child and Adolescent Psychiatry* 51, no. 8 (August 2012), doi: 10.1016/j.jaac.2012.05.018.

"*because you're just the chatty girl*": Morénike Giwa Onaiwu, interview with the author, 2018.

"*gender-based expectations*": Philippine Geelhand et al., "The Role of Gender in the Perception of Autism Symptom Severity and Future Behavioral Development," *Molecular Autism* 10, no. 1 (March 29, 2019), https://molecularautism.biomedcentral.com/articles/10.1186/s13229-019-0266-4#Tab1.

161 "*high functioning females with ASD*": Thomas W. Frazier et al., "Behavioral and Cognitive Characteristics of Females and Males with Autism in the Simons Simplex Collection," *Journal of the American Academy of Child and Adolescent Psychiatry* 53, no. 3 (2014), https://www.ncbi.nlm.nih.gov/pmc/articles/PMC3935179/.

surveyed 2,275 autistic people: Sander Begeer et al., "Sex Differences in the Timing of Identification Among Children and Adults with Autism Spectrum Disorders," *Journal of Autism and Developmental Disorders* 43, no. 5 (September 22, 2012), https://link.springer.com/article/10.1007/s10803-012-1656-z.

surveyed 2,568 children born in 1994: Ellen Giarelli et al., "Sex Differences in the Evaluation and Diagnosis of Autism Spectrum Disorders Among Children," *Disability and Health Journal* 3, no. 2 (2010), 107–16, https://pubmed.ncbi.nlm.nih.gov/21122776/.

women had higher camouflaging scores: Meng-Chuan Lai et al., "Quantifying and Exploring Camouflaging in Men and Women with Autism," *Autism* 21, no. 6 (November 29, 2016), https://journals.sagepub.com/doi/full/10.1177/1362361316671012.

"*there are few specialists*": M. Remi Yergeau, interview with the author, 2020.

162 *title of one of her books:* Liane Holliday Willey, *Pretending to Be Normal: Living with Asperger's Syndrome* (London: Jessica Kingsley, 2015).

"*We blend in easier because society makes excuses*": Liane Holliday Willey, interview with the author, 2015.

went undiagnosed until she was thirty: Ashia Ray, interview with the author, 2018.

163 "*My mom pretty much viewed me being autistic as her fault*": Eryn Star, interview with the author, 2018.

164 *surveyed fourteen women who were diagnosed with autism later in life:* Sarah Bargiela, Robyn Steward, and William Mandy, "The Experiences of Late-Diagnosed Women with Autism Spectrum Conditions: An Investigation of the Female Autism Phenotype," *Journal of Autism and Developmental Disorders* 46, no. 10 (July 25, 2016): 3281–94, https://link.springer.com/article/10.1007/s10803-016-2872-8.

"*feeling obliged*": Bargiela, Steward, and Mandy, "Experiences of Late-Diagnosed Women," 3288.

"*Whatever they told me*": Liane Holliday Willey, interview with the author, 2015.

165 "*I don't want it to happen to kids that come after me*": Liane Holliday Willey, interview with the author, 2015.

"*incompatible with how they wanted to live*": Bargiela, Steward, and Mandy, "Experiences of Late-Diagnosed Women," 3290.

"*The schema of the autistic male is sort of quirky and awkward*": Kris Harrison, interview with the author, 2018.

166 "*That realm of humor was off-limits*": Kris Harrison, interview with the author, 2019.

"*I feel like I'd be more myself in a man's body*": Kris Harrison, interview with the author, 2018.

"*socially expected skills*": Bargiela, Steward, and Mandy, "Experiences of Late-Diagnosed Women," 3289.

167 "*We can see that a lot of the social rules around gender*": Charlie Garcia-Spiegel, interview with the author, 2018.

"*gender variance was 7.59 times more common*": John F. Strang et al., "Increased Gender Variance in Autism Spectrum Disorders and Attention Deficit Hyperactivity Disorder," *Archives of Sexual Behavior* 43, no. 8 (2014): 1525–33, https://link.springer.com/article/10.1007%2Fs10508-014-0285-3.

"*behavior that falls outside of culturally defined norms associated with a specific gender*": Lisa K. Simons, Scott F. Leibowitz, and Marco A. Hidalgo, "Understanding Gender Variance in Children and Adolescents," *Pediatric Annals* 43, no. 6 (June 2014): 126–31, https://pubmed.ncbi.nlm.nih.gov/24972420/.

four times more likely to experience gender dysphoria: Elizabeth Hisle-Gorman et al., "Gender Dysphoria in Children with Autism Spectrum Disorder," *LGBT Health* 6, no. 3 (April 1, 2019), https://pubmed.ncbi.nlm.nih.gov/30920347/.

"*When we're forcibly distanced from social rules*": Charlie Garcia-Spiegel, interview with the author, 2018.

168 "*implied that I wasn't qualified to be at college*": Charlie Garcia-Spiegel, interview with the author, 2018.

"*autistic girls are hugely overrepresented*": J. K. Rowling, "J. K. Rowling Writes About Her Reasons for Speaking Out on Sex and Gender Issues," June 10, 2020, https://www.jkrowling.com/opinions/j-k-rowling-writes-about-her-reasons-for-speaking-out-on-sex-and-gender-issues/.

a noxious mix of transphobia and ableism: Kris Guin, "Autistic, Trans, and Betrayed by J. K. Rowling," *Thinking Person's Guide to Autism*, June 11, 2020, http://www.thinkingautismguide.com/2020/06/autistic-trans-and-betrayed-by-jk.html.

169 *less common sexual identities:* Laura Dattaro, "Gender and Sexuality in Autism, Explained," *Spectrum News,* September 18, 2020, https://www.spectrumnews.org/news/gender-and-sexuality-in-autism-explained/.

What Every Autistic Girl Wishes Her Parents Knew: *What Every Autistic Girl Wishes Her Parents Knew,* ed. Emily Paige Ballou, Kristina Thomas, and Sharon daVanport (Lincoln, NE: DragonBee Press, 2017).

discussed being autistic: Terry Gross, "Autism Spectrum Diagnosis Helped Comic Hannah Gadsby 'Be Kinder' to Herself," NPR, March 26, 2020, https://www.npr.org/2020/05/26/862081893/autism-spectrum-diagnosis-helped-comic-hannah-gadsby-be-kinder-to-herself.

170 *Cromer came out as autistic:* Anna Tingley, "'Everything's Gonna Be Okay' Star Discloses She's on Autism Spectrum," *Variety,* March 28, 2019, https://variety.com/2019/tv/news/kayla-cromer-austim-everythings-gonna-be-okay-freeform-1203174976/.

told me that in the initial storyboards: Erica Milsom, interview with the author, 2020.

171 *have hired consultants:* Hanh Nguyen, "'The Good Doctor' Upends Misconceptions About Sex and Relationships on the Spectrum," *Indiewire,* January 28, 2019, https://

www.indiewire.com/2019/01/the-good-doctor-sex-love-romance-xin-melissa-reiner-1202039446/.

further stigmatized autism and autistic people: Lindsey Bever, "How a 'Sesame Street' Muppet Became Embroiled in a Controversy over Autism," *Washington Post,* September 19, 2019, https://www.washingtonpost.com/health/2019/09/19/how-sesame-street-muppet-became-embroiled-controversy-over-autism/.

172 *"you need to know the interior life of the character":* Erica Milsom, interview with the author, 2020.

vocal parts were recorded at her house: Erica Milsom, interview with the author, 2020.

suggested she deserves the Nobel Prize: Damian Carrington, "Greta Thunberg Nominated for Nobel Peace Prize," *Guardian,* March 14, 2019, https://www.theguardian.com/world/2019/mar/14/greta-thunberg-nominated-nobel-peace-prize.

a reason she can be so strident: Charlotte Alter, Suyin Haynes, and Justin Worland, "Person of the Year: Greta Thunberg," *Time,* December 2019, https://time.com/person-of-the-year-2019-greta-thunberg/.

climate change in primary school: Greta Thunberg, "The Disarming Case to Act Right Now on Climate Change," Ted Talks, January 28, 2019, https://www.ted.com/talks/greta_thunberg_the_disarming_case_to_act_right_now_on_climate_change/transcript?language=en.

8. "Say It Loud"

174 *behavioral aide of Arnaldo Rios-Soto:* Aneri Pattani and Audrey Quinn, "What Happened Next to the Man with Autism Whose Aide Was Shot by Police," *Washington Post,* June 22, 2018, https://www.washingtonpost.com/news/to-your-health/wp/2018/06/22/what-happened-next-to-the-man-with-autism-whose-aide-was-shot-by-police/.

bleeding for twenty minutes: Francisco Alvarado, Michael E. Miller, and Mark Berman, "North Miami police shoot black man who said his hands were raised while he tried to help autistic group-home resident," *Washington Post,* July 21, 2016, https://www.washingtonpost.com/news/morning-mix/wp/2016/07/21/fla-police-shoot-black-man-with-his-hands-up-as-he-tries-to-help-autistic-patient/.

175 *was shot and killed by a Hispanic police officer:* Emily Shapiro and Julia Jacobo, "Minnesota Officer Fired from Police Force After Acquittal in Philando Castile Shooting," *ABC News,* June 16, 2017, https://abcnews.go.com/US/minnesota-officer-found-guilty-fatal-shooting-philando-castile/story?id=48003144.

justified the shooting: Alan Gomez, "Police Union Says Officers Accidentally Shot Miami Man," *USA Today,* July 21, 2016, https://www.usatoday.com/story/news/2016/07/21/north-miami-police-shooting-accidental-union-says/87414898/.

"perpetual child": Maxfield Sparrow, "It's Time to Prepare the World for Your Child," *Thinking Person's Guide to Autism,* December 20, 2018, http://www.thinkingautismguide.com/2018/12/its-time-to-prepare-world-for-your-child.html.

176 *"No one wants to be looked at":* A. J. Link, interview with the author, 2019.

177 *nine of them were from Anglo-Saxon families:* Kanner, "Autistic Disturbances of Affective Contact," 248.

represented either in Who's Who in America*:* Steve Silberman, "The Invisibility of Black Autism," *Undark,* May 5, 2016, https://undark.org/2016/05/17/invisibility-black-autism/.

one of those cold and unloving parents: David E. Simpson, *Refrigerator Mothers,* Kartemquin Films, March 28, 2003, https://kartemquin.vhx.tv/products/refrigerator-mothers.

"*had 2.6 times the odds of receiving some other diagnosis*": David S. Mandell et al., "Disparities in Diagnoses Received Prior to a Diagnosis of Autism Spectrum Disorder," *Journal of Autism and Developmental Disorders* 37, no. 9 (December 8, 2006), https://www.ncbi.nlm.nih.gov/pmc/articles/PMC2861330/.

said that the gap between white children diagnosed with autism: "Community Report on Autism 2020," Centers for Disease Control and Prevention, 2020, https://www.cdc.gov/ncbddd/autism/addm-community-report/documents/addm-community-report-2020-h.pdf.

white and Black children are 1.2 times more likely to be diagnosed than Hispanic children: Community Report on Autism 2020."

178 *they required more time in treatment before they were diagnosed:* David S. Mandell et al., "Race Differences in the Age at Diagnosis Among Medicaid-Eligible Children with Autism," *Journal of the American Academy of Child and Adolescent Psychiatry* 41, no. 12 (2002), https://pubmed.ncbi.nlm.nih.gov/12447031/.

the CDC confirmed that Black and Hispanic children get diagnosed later: Centers for Disease Control and Prevention, "Community Report on Autism 2020," 15.

"*I would often skip exams or not turn in papers*": A. J. Link, interview with the author, 2019.

179 *5.1 times more likely to be diagnosed with adjustment disorder:* Mandell et al., "Disparities in Diagnoses."

a diagnosis of conduct disorder: Mandell et al., "Disparities in Diagnoses."

"difficulty following rules, respecting the rights of others": "Conduct Disorder," *American Academy of Child and Adolescent Psychiatry,* no. 33, updated June 2018, https://www.aacap.org/AACAP/Families_and_Youth/Facts_for_Families/FFF-Guide/Conduct-Disorder-033.aspx.

might also diagnose Black children with oppositional defiant disorder: Mandell et al., "Disparities in Diagnoses."

"*I feel that I experienced a lot of disproportionate disciplines*": Finn Gardiner, interview with the author, 2018.

180 "*I got in fights with other kids a lot*": Sara Luterman, interview with the author, 2019.

Black parents reported fewer concerns: Meghan Rose Donohue et al., "Race Influences Parent Report of Concerns About Symptoms of Autism Spectrum Disorder," *Autism* 23, no. 1 (November 3, 2017): 106, https://journals.sagepub.com/doi/pdf/10.1177/1362361317722030.

181 "*cultural norms surrounding discussion of ASD*": Donohue et al., "Race Influences Parent Report," 107.

"*the combined effects of practices*": Kimberlé Crenshaw, "Demarginalizing the Intersection of Race and Sex: A Black Feminist Critique of Antidiscrimination Doctrine, Feminist Theory and Antiracist Politics," *University of Chicago Legal Forum* 1989, no. 1 (1989): 149, https://chicagounbound.uchicago.edu/cgi/viewcontent.cgi?article=1052&context=uclf.

182 "*my own people don't understand me*": Morénike Giwa Onaiwu, interview with the author, 2018.

"*The first way I got into the in-crowd*": Timotheus Gordon, interview with the author, 2018.

184 *30.4 percent offered general developmental screenings:* Katharine E. Zuckerman et al., "Pediatrician Identification of Latino Children at Risk for Autism Spectrum Disorder," *Pediatrics* 132, no. 3 (August 8, 2013): 448, https://pediatrics.aappublications.org/content/pediatrics/132/3/445.full.pdf.

"*it's such a confusing situation for kids*": Katharine Zuckerman, interview with the author, 2018.

185 "*There's a reason I didn't get that diagnosis when I was younger*": Arianne Garcia, interview with the author, 2018.

"*I had trouble nursing*": Debra Garcia, interview with the author, 2019.

186 "*parent knowledge about ASD*": Katharine E. Zuckerman et al., "Disparities in Diagnosis and Treatment of Autism in Latino and Non-Latino White Families," *Pediatrics* 139, no. 5 (April 24, 2017), https://pediatrics.aappublications.org/content/139/5/e20163010.

187 "*I'm not saying most Latino parents think that*": Katharine Zuckerman, interview with the author, 2018.

calls for the arrests of the officers who killed her: Aja Romano, "'Arrest the cops who killed Breonna Taylor': The power and the peril of a catchphrase," *Vox,* August 10, 2020, https://www.vox.com/21327268/breonna-taylor-say-her-name-meme-hashtag.

video was captured of George Floyd's murder: Evan Hill et al., "How George Floyd Was Killed in Police Custody," *New York Times,* May 31, 2020, https://www.nytimes.com/2020/05/31/us/george-floyd-investigation.html.

Jacob Blake's shooting in Kenosha, Wisconsin: Christina Morales, "What We Know About the Shooting of Jacob Blake," *New York Times,* September 10, 2020, https://www.nytimes.com/article/jacob-blake-shooting-kenosha.html.

188 *Black Lives Matter:* Michael Tesler, "Support for Black Lives Matter Surged During Protests, But Is Waning Among White Americans," *FiveThirtyEight,* August 19, 2020, https://fivethirtyeight.com/features/support-for-black-lives-matter-surged-during-protests-but-is-waning-among-white-americans/.

5 percent of them had been arrested: Julianna Rava et al., "The Prevalence and Correlates of Involvement in the Criminal Justice System Among Youth on the Autism Spectrum," *Journal of Autism Developmental* 47, no. 2 (February 2017): 340–46, https://pubmed.ncbi.nlm.nih.gov/27844248/.

Chicago suburb of Calumet City: Danelene Powell-Watts (as Special Administrator) Appeal from the Estate of Stephon Edward Watts, Deceased v. City of Calumet City, a Municipal Corporation; William Coffey; and Robert Hynek, https://courts.illinois.gov/r23_orders/AppellateCourt/2016/1stDistrict/1151973_R23.pdf.

ruled in favor of the police: Elvia Malagon, "Court Rules in Favor of Cal City Cops in Fatal Shooting of Autistic Teen," *Times of Northwest Indiana,* July 7, 2016, https://www.nwitimes.com/news/local/illinois/court-rules-in-favor-of-cal-city-cops-in-fatal-shooting-of-autistic-teen/article_b6ccccb4-ec61-5f43-ab58-bd65750328d1.html.

dashcam footage showed: Don Babwin, "Video Shows Chicago Cop Shooting Unarmed Black Autistic Teen," *Daily Herald,* October 17, 2018, https://www.dailyherald.com/article/20181017/news/310179954.

Ricardo Hayes: Todd Feurer, "Aldermen Back $2.25 Million Payment to Unarmed Autistic Man Shot by Police; Two Other Lawsuit Settlements," CBS, May 18, 2020, https://chicago.cbslocal.com/2020/05/18/aldermen-back-2-25-million-payment-to-unarmed-autistic-man-shot-by-police-two-other-lawsuit-settlements/.

189 *not being respectful:* Steve Silberman, "The Police Need to Understand Autism," *New*

York Times, September 19, 2017, https://www.nytimes.com/2017/09/19/opinion/police-autism-understanding.html.

often perceived to be older than they are: Phillip Atiba Goff et al., "The Essence of Innocence: Consequences of Dehumanizing Black Children," *Journal of Personality and Social Psychology* 106, no. 4 (2014): 526–45, https://www.apa.org/pubs/journals/releases/psp-a0035663.pdf.

"*when our children have episodes":* Eric Garcia, "What It Feels Like to Be an Autistic Person of Color in the Eyes of the Police," *Daily Beast,* July 25, 2016, https://www.thedailybeast.com/what-it-feels-like-to-be-an-autistic-person-of-color-in-the-eyes-of-the-police.

the three white officers were initially cleared: Allison Sherry, "Feds Have a High Bar for Charging Aurora Police in Elijah McClain's Death—but a Case in Westminster Shows It Can Happen," *CPR News,* July 30, 2020, https://www.cpr.org/2020/07/30/federal-investigation-prosecuting-aurora-police-elijah-mcclain-death-westminster-curtis-lee-arganbright/.

from getting iced tea for his brother: Sonia Gutierrez, "'Let's Honor His Spirit': Friend of Elijah McClain Said Allowing Officer in Photo to Resign Isn't Justice," KTVB7, July 2, 2020, https://www.ktvb.com/article/news/local/officers-resignation-worries-elijah-mcclains-friends/73-7798c217-8456-44b4-9a46-1c691fe2fc67.

"waving his arms around": Ellie Hall, "Elijah McClain Died in Police Custody in August. Millions of People Are Now Demanding Justice," *BuzzFeed News,* June 24, 2020, https://www.buzzfeednews.com/article/ellievhall/elijah-mcclain-aurora-death-police-investigation-demand.

"I'm an introvert": Knez Walker et al., "What Happened to Elijah McClain? Protests Help Bring New Attention to His Death," *ABC News,* June 30, 2020, https://abcnews.go.com/US/happened-elijah-mcclain-protests-bring-attention-death/story?id=71523476.

put him in a carotid hold: "Aurora Police Announce No Charges in Death of Elijah McClain Following Arrest," CBS Denver, November 22, 2019, https://denver.cbslocal.com/2019/11/22/aurora-police-elijah-mcclain-body-camera-autopsy/.

190 *being "different" makes them a target:* Jackie Spinner, "Elijah McClain's Final Words Haunt Me as the Parent of a Child Who Is 'Different,'" *Washington Post,* June 29, 2020, https://www.washingtonpost.com/opinions/2020/06/29/elijah-mcclains-last-words-haunt-me-could-that-happen-my-son/.

use virtual reality to train police: Michael Balsamo, "Virtual Reality Helps Police in Dealing with Autistic People," Associated Press, May 24, 2019, https://apnews.com/article/5e4c1aba96b54d7c9c791d3d83d1bfea.

Pennsylvania and New Jersey have passed similar bills: Hannah Furfaro, "Why Police Need Training to Interact with People on the Spectrum," *Spectrum News,* June 6, 2018, https://www.spectrumnews.org/features/deep-dive/police-need-training-interact-people-spectrum/.

23 percent of respondents to a survey: Edward Kelly and Connie Hassett-Walker, "The Training of New Jersey Emergency Service First Responders in Autism Awareness," *Police Practice and Research* 17, no. 6 (January 4, 2016), https://www.ncbi.nlm.nih.gov/pmc/articles/PMC5363970/.

mandate training for new police officers: Tara Malone, "Autism Training Helps Police Tailor Response," *Chicago Tribune,* July 21, 2008, https://www.chicagotribune.com/news/ct-xpm-2008-07-21-0807210022-story.html.

defunding police departments: Shayla Love, "Police Are the First to Respond to Mental Health Crises. They Shouldn't Be," *Vice,* June 23, 2020, https://www.vice.com/en/article/3azkeb/police-are-the-first-to-respond-to-mental-health-crises-they-shouldnt-be.

193 "*I realize [the police officer]*": Morénike Giwa Onaiwu, interview with the author, 2018.

194 "*listen to three or four minutes and it kind of just puts me at ease*": A. J. Link, interview with the author, 2019.

"*I feel like an anthropologist on Mars*": Oliver Sacks, "An Anthropologist on Mars," *New Yorker,* December 27, 1993, https://www.newyorker.com/magazine/1993/12/27/anthropologist-mars.

195 *adults with Hispanic ancestry:* Mark Hugo Lopez, Ana Gonzalez-Barrera, and Gustavo López, "Hispanic Identity Fades Across Generations as Immigrant Connections Fall Away," Pew Research Center, December 20, 2017, https://www.pewresearch.org/hispanic/2017/12/20/hispanic-identity-fades-across-generations-as-immigrant-connections-fall-away/.

9. "Till the Next Episode"

198 "*take [him] for every penny you still don't have*": Tim Mak, "Listen: How Michael Cohen Protects Trump by Making Legal Threats," NPR, May 31, 2018, https://www.npr.org/2018/05/31/615843930/listen-how-michael-cohen-protects-trump-by-making-legal-threats.

199 *a columnist at* National Journal: Ron Fournier, "How Two Presidents Helped Me Deal with Love, Guilt, and Fatherhood," *National Journal,* November 29, 2012, https://www.theatlantic.com/politics/archive/2012/11/how-two-presidents-helped-me-deal-with-love-guilt-and-fatherhood/461733/.

"*The Geek Syndrome*": Steve Silberman, "The Geek Syndrome," *Wired,* December 12, 2001, https://www.wired.com/2001/12/aspergers/.

graph shaped like a hockey stick: Dominic Basulto, "We Need a Better Explanation for the Surge in Autism," *Washington Post,* April 10, 2014, https://www.washingtonpost.com/news/innovations/wp/2014/04/10/we-need-a-better-explanation-for-the-surge-in-autism/.

200 *homosexuality was even listed in the* DSM: Jack Drescher, "Out of *DSM*: Depathologizing Homosexuality," *Behavioral Sciences* 5, no. 4 (December 4, 2015), https://www.ncbi.nlm.nih.gov/pmc/articles/PMC4695779/.

lives with his husband, Keith Karraker: Jenny Turner, "The Man Who Wants Us to Embrace Autism," *Guardian,* August 29, 2015, https://www.theguardian.com/lifeandstyle/2015/aug/29/autism-spectrum-steve-silberman-neurotribes-legacy-autism-people-think-differently.

"*If you think that homosexuality or autism is rare*": Steve Silberman, interview with the author, 2020.

placed all autistic people under the larger diagnosis: Turner, "The Man Who Wants Us to Embrace Autism."

experiences of autistic people of color: Lydia X. Z. Brown, Morénike Giwa Onaiwu, and E. Ashkenazy, eds, *All the Weight of Our Dreams: On Living Racialized Autism* (Lincoln, NE: DragonBee Press, 2017).

201 "*autism isn't an illness*": Barry Prizant with Tom Fields-Meyer, *Uniquely Human: A Different Way of Seeing Autism* (New York: Simon and Schuster, 2015), 4.

surveyed 315 articles from the Washington Post: Noa Lewin and Nameera Akhtar, "Neurodiversity and Deficit Perspectives in the *Washington Post*'s Coverage of Autism," *Disability and Society* (May 2020), https://www.tandfonline.com/doi/abs/10.1080/09687599.2020.1751073.

write a piece on autism, sex, and consent: Garcia, "Autistic Men Don't Always Understand Consent."

to amplify autistic: Lydia X. Z. Brown, "Autistic Young People Deserve Serious Respect and Attention—Not Dismissal as the Pawns of Others," *Washington Post,* December 14, 2019, https://www.washingtonpost.com/outlook/2019/12/14/autistic-young-people-deserve-serious-respect-attention-not-dismissal-pawns-others/.

otherwise disabled voices: Rebecca Cokley, "Calling Trump Unwell Doesn't Hurt Trump. It Hurts Disabled People," *Washington Post,* June 16, 2020, https://www.washingtonpost.com/outlook/2020/06/16/mock-trump-hurts-disabled/.

202 *Hillary Clinton became the first presidential candidate:* Eric Garcia, "Autism Advocates Cautiously Optimistic on Clinton Proposal," *Roll Call,* January 6, 2016, https://www.rollcall.com/2016/01/06/autism-advocates-cautiously-optimistic-on-clinton-proposal/.

common rhetoric about "combating": Hillary Clinton, "Remarks at the Autism Event with Sally Pederson in Sioux City, Iowa," *American Presidency Project,* November, 24, 2007, https://www.presidency.ucsb.edu/documents/remarks-the-autism-event-with-sally-pederson-sioux-city-iowa.

"money into finding causes and cures of diseases": Hillary Clinton, "Clinton Campaign Event," C-SPAN, January 7, 2008, https://www.c-span.org/video/?203432-1/clinton-campaign-event.

But her policy approach shifted drastically by 2016: Hillary Clinton, "Autism," Hillaryclinton.com, 2016, https://www.hillaryclinton.com/issues/autism/.

her campaign plan read like a wish list: Eric Garcia, "Autism Advocates Cautiously Optimistic."

"*There is a huge physical and cognitive cost*": Julia Bascom, interview with the author, 2020.

203 *gained an incredible amount of political capital:* Sady Doyle, "If You Celebrated the Health Care Vote Last Week, You Should Probably Thank a Disability Activist," *Elle,* August, 1, 2017, https://www.elle.com/culture/career-politics/news/a47075/disability-activist-health-care/.

PLEASE SAVE OUR HEALTH CARE: Julia Bascom, correspondence with the author, 2020.

Bascom said she consulted with candidates: Julia Bascom, interview with the author, 2020.

204 "*political, financial, and nonprofit infrastructure*": Ari Ne'eman, interview with the author, 2019.

"*primaries are about articulating a vision*": Julia Bascom, interview with the author, 2020.

many Democrats thought Biden: Ronald Brownstein, "The Democratic Debate Over Winning Back Trump's Base," *Atlantic,* May 2, 2019, https://www.theatlantic.com/politics/archive/2019/05/joe-bidens-bid-white-working-class-vote/588613/.

she spoke with Biden's team during the primary: Julia Bascom, interview with the author, 2020.

prompting the hashtag #AccessToJoe on Twitter: Sara Luterman, "Biden's Disability Policy Plan Is Surprisingly Good," *Nation,* June 1, 2020, https://www.thenation.com/article/politics/biden-disability-policy-plan/.

205 *finally released a disability policy platform:* Joe Biden, "The Biden Plan for Full Participation and Equality for People with Disabilities," June 2020, https://joebiden.com/disabilities/#.

everybody has some form of disability: Luterman, "Biden's Disability Policy Plan Is Surprisingly Good."

206 *died at a higher rate:* Joseph Shapiro, "COVID-19 Infections and Deaths Are Higher Among Those with Intellectual Disabilities," NPR, June 9, 2020, https://www.npr.org/2020/06/09/872401607/covid-19-infections-and-deaths-are-higher-among-those-with-intellectual-disabili.

"*I didn't want to claim that title*": Molly Doris-Pierce, interview with the author, 2020.

"*Nothing about us without us*": Ayanna Pressley, reporting by the author, 2020.

the slogan of the self-advocacy movement for disability rights: Charlton, *Nothing About Us Without Us.*

207 *Lara Lea Trump did highlight that under his administration:* Lara Trump, "Lara Trump Speaks on Behalf of the Trump Campaign during AAPD REV UP 2020 Summit," YouTube, June 23, 2020, https://www.youtube.com/watch?v=tQJ9NKrIhLI.

the first Democratic presidential candidate: Sarah Kim, "President Hopeful Kamala Harris First 2020 Candidate to Announce Exclusive Plan for People with Disabilities," *Forbes,* August 30, 2019, https://www.forbes.com/sites/sarahkim/2019/08/30/kamala-harris-disability-plan/#7c8674561234.

"Institutions for Mental Disease Exclusion": "Kamala Harris Unveils New Mental Health Care Plan," *Democracy in Action,* November 25, 2019, https://www.democracyinaction.us/2020/harris/harrispolicy112519mentalhealth.html.

The IMD exclusion prevents Medicaid: Alison Mitchell, "Medicaid's Institutions for Mental Disease (IMD) Exclusion," Congressional Research Service, https://fas.org/sgp/crs/misc/IF10222.pdf.

repealing the exclusion: Sara Luterman, "How Kamala Harris's Mental Health Plan Could Hurt the Most Vulnerable," Vox.com, November 27, 2019, https://www.vox.com/first-person/2019/11/27/20985430/kamala-harris-mental-health-plan.

planning on reworking her mental-health plan: Reporting by the author, 2020.

"*similar to the drive to view autism*": John Hendrickson, "What Joe Biden Can't Bring Himself to Say," *Atlantic,* January/February 2020, https://www.theatlantic.com/magazine/archive/2020/01/joe-biden-stutter-profile/602401/.

pin that supports the Autism Alliance in Michigan: Tim Alberta, Twitter, December 20, 2019, https://twitter.com/TimAlberta/status/1208118246584541184.

Abby Abrams at TIME: Abby Abrams, "Voter Turnout Surged Among People With Disabilities Last Year. Activists Want to Make Sure That Continues in 2020," *Time,* July 10, 2019, https://time.com/5622652/disability-voter-turnout-2020/.

Maggie Astor: Maggie Astor, "Elizabeth Warren Opens a New Front in Disability Policy," *New York Times,* January 10, 2020, https://www.nytimes.com/2020/01/10/us/politics/elizabeth-warren-disability-plan.html.

208 *a disabled father:* David M. Perry, "My Coronavirus Prep Includes Protecting My

Disabled Child," *Washington Post,* March 4, 2020, https://www.washingtonpost.com/outlook/2020/03/04/coronavirus-prep-disabled-children/.

when I saw a tweet from right-wing provocateur Ryan Saavedra: Ryan Saavedra, Twitter, September, 6, 2019, https://twitter.com/RealSaavedra/status/1170082834377105410.

"*I hope @KamalaHarris enjoys losing the disability vote":* Julia Bascom, Twitter, September, 7, 2019, https://twitter.com/JustStimming/status/1170392597178789889.

he would be up to handling this story correctly: Zack Budryk, "Harris Apologizes and Says She Didn't Hear Questioner Call Trump Policies 'Mentally Retarded,'" *Hill,* September 7, 2019, https://thehill.com/homenews/campaign/460385-harris-says-she-didnt-hear-questioner-call-trump-policies-mentally-retarded.

209 *quote autistic self-advocates like Ari Ne'eman:* Sheri Fink, "U.S. Civil Rights Office Rejects Rationing Medical Care Based on Disability, Age," *New York Times,* March 28, 2020, https://www.nytimes.com/2020/03/28/us/coronavirus-disabilities-rationing-ventilators-triage.html.

cover parents like Shannon Des Roches Rosa: Rose Eveleth, "Beyond the Smiley-Face Pain Scale," *Atlantic,* January, 7, 2015, https://www.theatlantic.com/health/archive/2015/01/beyond-the-smiley-face-pain-scale/384049/.

210 *ban on recognizing same-sex marriages:* Eyder Peralta, "Court Overturns DOMA, Sidesteps Broad Gay Marriage Ruling," June 26, 2013, https://www.npr.org/sections/thetwo-way/2013/06/26/195857796/supreme-court-strikes-down-defense-of-marriage-act.

"*It's extraordinary to me that today":* Roberta Kaplan, interview with the author, 2020.

212 "*So, I was diagnosed at twelve":* Ari Ne'eman, interview with the author, 2019.

214 "*What I ask for the Negro is not benevolence";* Frederick Douglass, "What the Black Man Wants," 1865, http://utc.iath.virginia.edu/africam/afspfdat.html.

Epilogue

217 *Mel Baggs died:* Harrison Smith, "Mel Baggs, Influential Blogger on Disability and Autism, Dies at 39," *Washington Post,* April 29, 2020, https://www.washingtonpost.com/local/obituaries/mel-baggs-influential-blogger-on-disability-and-autism-dies-at-39/2020/04/29/bbb0fdd2-8a24-11ea-ac8a-fe9b8088e101_story.html.

218 "*I find it very interesting that":* Mel Baggs, "In My Language," YouTube, January 14, 2007, https://www.youtube.com/watch?v=JnylM1hI2jc.

finally landed a job after years of searching: Andrew Savicki, correspondence with the author, 2020.

219 "*There are people being tortured":* Mel Baggs, "In My Language."

Afterword

221 *preached how her autism is an asset:* D. E. Elizabeth, "Greta Thunberg on Her Autism Diagnosis and Climate Activism," *Teen Vogue,* September 26, 2021, https://www.teenvogue.com/story/greta-thunberg-autism-diagnosis-climate-activism.

autistic people were far more vulnerable to die from COVID-19 than neurotypical people: Katie Camero, "COVID-19 Kills Children with Intellectual Disabilities at Higher Rates. Here's Why," *Miami Herald,* July 19, 2020, https://www.miamiherald.com/news/coronavirus/article243470691.html.

especially those in congregate care facilities: Joseph Shapiro, "COVID-19 Infections and Deaths Are Higher Among Those with Intellectual Disabilities," June 9, 2020, NPR, https://www.npr.org/2020/06/09/872401607/covid-19-infections-and-deaths-are-higher-among-those-with-intellectual-disabili.

222 *whopping $400 billion on home- and community-based:* The White House, "Fact Sheet: The American Jobs Plan," March 31, 2021, https://www.whitehouse.gov/briefing-room/statements-releases/2021/03/31/fact-sheet-the-american-jobs-plan/.

split in half: Jim Tankersley, "Biden Team Prepares $3 Trillion in New Spending for the Economy." *New York Times,* March 22, 2021, https://www.nytimes.com/2021/03/22/business/biden-infrastructure-spending.html.

to pass a "hard" infrastructure: Manu Raju, Lauren Fox, and Ted Barrett, "Biden's Bipartisan Infrastructure Deal Could Face Key Senate GOP Defections," CNN, July 12, 2021, https://www.cnn.com/2021/07/12/politics/biden-infrastructure-senate-republican-opposition/index.html.

in November of 2021: Josh Boak and Colleen Long, "Biden Signs $1T Infrastructure Deal with Bipartisan Crowd," Associated Press, https://apnews.com/article/joe-biden-congress-infrastructure-bill-signing-b5b8cca843133de060778f049861b144.

which was reduced to $150 billion: The White House, "President Biden Announces the Build Back Better Framework," October 28, 2021, https://www.whitehouse.gov/briefing-room/statements-releases/2021/10/28/president-biden-announces-the-build-back-better-framework/.

Joe Manchin of West Virginia kill the legislation: Ronn Blitzer, "Manchin Says He 'Cannot Vote' for Build Back Better: 'I've Done Everything Humanly Possible.'" Fox News, December 19, 2021, https://www.foxnews.com/politics/manchin-says-he-cannot-vote-for-build-back-better-ive-done-everything-humanly-possible.

223 *vaccines on a QAnon Twitch stream:* Alex Kaplan, "Anti-Vax Influencers Are Using QAnon Shows to Spread Misinformation about COVID-19 Vaccines," Media Matters for America, May 11, 2021, https://www.mediamatters.org/coronavirus-covid-19/anti-vax-influencers-are-using-qanon-shows-spread-misinformation-about-covid.

lawyer turned anti-vaccine activist: Jonathan Jarry, "The Anti-Vaccine Propaganda of Robert F. Kennedy, Jr.," Office for Science and Society, April 16, 2021, https://www.mcgill.ca/oss/article/covid-19-health-pseudoscience/anti-vaccine-propaganda-robert-f-kennedy-jr.

linked to the COVID-19 vaccine: Olga Robinson, BBC Monitoring, "Instagram Bans Robert F. Kennedy Jr. over COVID Vaccine Posts," BBC, February 11, 2021, https://www.bbc.com/news/world-us-canada-56021904.

abuses at Willowbrook State School in New York: Minnesota Governor's Council on Developmental Disabilities, "Robert Kennedy Visiting Institutions in NY," YouTube, November 12, 2020, https://www.youtube.com/watch?v=iEwFXfy5EV0.

architect of the Americans with Disabilities Act: "Ending Segregation and Discrimination Against Disabled Americans," Edward M. Kennedy Institute, 2011, http://www.tedkennedy.org/ownwords/event/disabilities.html.

a documentary lionizing Wakefield: Rebecca Robbins, "We Watched the Movie *Vaxxed* so You Don't Have to," *Stat News,* April 1, 2016, https://www.statnews.com/2016/04/01/vaxxed-autism-movie-review/.

small rally on January 6, 2021, in Washington: Curt Devine and Drew Griffin, "Leaders of the Anti-Vaccine Movement Used 'Stop the Steal' Crusade to Advance Their Own Conspiracy Theories," CNN, February 5, 2021, https://www.cnn.com/2021/02/04/politics/anti-vaxxers-stop-the-steal-invs/index.html.

to borrow from Lincoln: Abraham Lincoln, The Gettysburg Address, 1863, transcript from Cornell University Library, https://rmc.library.cornell.edu/gettysburg/good_cause/transcript.htm.

Trumpism created a singularity: Anna Merlan, "The Conspiracy Singularity Has Arrived," *Vice,* July 11, 2020, https://www.vice.com/en/article/v7gz53/the-conspiracy-singularity-has-arrived.

"QAnon Shaman" for going shirtless: Hannah Rabinowitz and Katelyn Polantz, "'QAnon Shaman' Jacob Chansley Sentenced to 41 Months in Prison for Role in US Capitol Riot," CNN, November 17, 2021, https://www.cnn.com/2021/11/17/politics/jacob-chansley-qanon-shaman-january-6-sentencing/index.html.

224 *"they're on the goddamn spectrum":* Matt Shuham, "Capitol Rioters' 'Trump Defense' Comes Up Again and Again. Will It Make a Difference?" *Talking Points Memo,* May 18, 2021, https://talkingpointsmemo.com/news/capitol-rioters-trump-defense-comes-up-again-and-again-will-it-make-a-difference.

"brainwashed": Adam Wallis, "Climate Change Denier Meat Loaf Says Greta Thunberg Is 'Brainwashed.'" *Global News,* January 6, 2020, https://globalnews.ca/news/6371945/meat-loaf-calls-greta-thunberg-brainwashed/.

"marketing gimmick": Donald J. Trump Jr., Twitter, December 11, 2019, https://twitter.com/DonaldJTrumpJr/status/1204768478722433024.

society doesn't really know how to handle autistic girls: Anna North, "Attacks on Greta Thunberg Expose the Stigma Autistic Girls Face," *Vox,* December 12, 2021, https://www.vox.com/identities/2019/9/24/20881837/greta-thunberg-person-of-the-year-trump.

is often seen as a threat: Rebecca Traister, "And You Thought Trump Voters Were Mad," *New York,* September 17, 2018, https://www.thecut.com/2018/09/rebecca-traister-good-and-mad-book-excerpt.html.

he was sentenced to prison: Rabinowitz and Polants, "'QAnon Shaman' Jacob Chansley Sentenced to 41 Months in Prison for Role in US Capitol Riot."

225 *David M. Perry asked me:* David M. Perry, "Review/Interview: *We're Not Broken* by Eric Garcia," *Patreon,* July 30, 2021. https://www.patreon.com/posts/review-interview-54068602.

and her neurotypical sister: Clem Bastow, "Sia's Film *Music* Misrepresents Autistic People. It Could Also Do Us Damage," January 26, 2021, https://www.theguardian.com/film/2021/jan/27/sias-film-music-misrepresents-autistic-people-it-could-also-do-us-damage.

nothing to move the plot along: David Ehrlich, "*Music* Review: Sia's Directorial Debut Is a Staggeringly Tone-Deaf Display of Magical Thinking," *Indiewire,* February 10, 2021, https://www.indiewire.com/2021/02/music-review-sia-autism-movie-maddie-ziegler-1234615917/.

*"*Rain Man, *the musical, but with girls":* Shirley Halperin, "Sia Talks Directing Her First Feature Film *Music,*" *Variety,* October 29, 2020, https://www.youtube.com/watch?v=SIVppt0YPio.

"nonverbal on the spectrum": Zack Sharf, "Sia Defends Casting Maddie Ziegler as Autistic Teen in *Music,* Official Trailer Debuts—Watch," *Indiewire,* December 16, 2020,

https://www.indiewire.com/2020/12/sia-defends-casting-maddie-ziegler-autistic-music-trailer-1234605197/.

"unpleasant and stressful": Sharf, "Sia Defends Casting Maddie Ziegler as Autistic Teen in *Music,* Official Trailer Debuts—Watch."

"Maybe you're just a bad actor": David Oliver, "'I Should Have Just Shut up': Sia Walks Back Comments after Being Criticized for Ableism," *USA Today,* December 20, 2020, https://www.usatoday.com/story/entertainment/celebrities/2020/12/21/sia-walks-back-autism-casting-comments-after-criticism-ableism/3989553001/.

"watch my film before you judge it?": Leonie Cooper, "*Music* Review: Sia's Clumsy Autism Drama Hits All the Wrong Notes," *NME,* February 16, 2021, https://www.nme.com/reviews/film-reviews/music-review-sia-2881672.

226 *Ziegler's character:* Joey Nolfi, "Sia Says *Music* Movie Will Include Warning Label, Cuts Controversial Restraint Scenes," *Entertainment Weekly,* February 4, 2021, https://ew.com/movies/sia-music-controversy-deletes-twitter/.

when used to stop meltdowns: Renee Fabian, "Sia Addresses Depictions of Lethal Restraint During Autistic Meltdown in *Music,*" *Mighty,* January 25, 2021, https://themighty.com/2021/01/sia-music-prone-restraint-autistic-meltdown/.

she deactivated her Twitter account: Nolfi, "Sia Says *Music* Movie Will Include Warning Label, Cuts Controversial Restraint Scenes."

"love letter": Shona Leigh Pope, "Sia's *Music* Is Not 'a Love Letter to the Autistic Community': It's Another Unrealistic Portrayal of ASD," *Film,* February 25, 2021, https://www.thefilmagazine.com/sia-music-movie-unrealistic-portrayal-of-asd/.

Index